SOMETHING HIDDEN IN THE RANGES

SOMETHING HIDDEN IN THE RANGES

The Secret Life of Mountain Ecosystems

ELLEN WOHL

Oregon State University Press   Corvallis

The John and Shirley Byrne Fund for Books on Nature and the Environment provides generous support that helps make publication of this and other Oregon State University Press books possible.

Library of Congress Cataloging-in-Publication Data

Names: Wohl, Ellen E., 1962– author.
Title: Something hidden in the ranges : the secret life of mountain ecosystems / Ellen Wohl.
Description: Corvallis : Oregon State University Press, [2021] | Includes bibliographical references and index.
Identifiers: LCCN 2021005697 | ISBN 9780870711053 (trade paperback) | ISBN 9780870711060 (ebook)
Subjects: LCSH: Mountain ecology.
Classification: LCC QH541.5.M65 W64 2021 | DDC 577.5/3--dc23
LC record available at https://lccn.loc.gov/2021005697

 Oregon State University
OSU Press

Oregon State University Press
121 The Valley Library
Corvallis OR 97331-4501
541-737-3166 • fax 541-737-3170
www.osupress.oregonstate.edu

CONTENTS

ACKNOWLEDGMENTS

This book benefited from review comments by Jon Luoma, Fred Swanson, and SueEllen Campbell. I appreciate the many research permits granted to me by the National Park Service during the past thirty years, which greatly facilitated my access to Rocky Mountain National Park. I relied heavily on the second edition of the informative book *Mammals of Colorado* by David Armstrong, James Fitzgerald, and Carron Meaney for my descriptions of the ecology of mammals in Rocky Mountain National Park. And I relied on dozens of peer-reviewed scientific articles and books summarizing original research in and related to the national park: I gratefully acknowledge the dedication, perseverance, and passion of the members of the scientific community who conducted this research. Finally, I appreciate the willingness of Kim Hogeland and Oregon State University Press to publish a book about Colorado.

PROLOGUE

It's early August 2020, and I've come to Rocky Mountain National Park for an afternoon of reprieve from the chaotic news of this pandemic year. I'm sitting next to a snippet of creek at the east end of Horseshoe Park, inside one of the areas where hungry elk are fenced out so that shrubs and trees can flourish. What little water is left in Fall River this late in the summer flows quietly in front of me, and a thicket of willows and alder stands behind my back. The river here has cut itself a deep channel—the banks are maybe three feet high—in what was clearly not all that long ago, certainly during my lifetime, a meadow-wide home to beaver colonies. And indeed, I've chosen this spot today because at least a couple of those alluring creatures have moved back in just downstream and across from me, as I learned two summers ago from my friend Ellen Wohl. I can't tell whether they've tried to build a dam across the water, but they've erected a big pile of cut branches on the bank opposite me, so I'm keeping an eye on the water in case a beaver ventures out of hiding to swim upstream past me: unlikely at this time of the day, high afternoon, but not impossible.

Actually, I've been coming to this tiny retreat often over the last decade or so, since I discovered it just steps from the road but hidden from the sight and most of the sound of the park's busy tourist traffic. I've spent hours here, sometimes holding a book or notepad, often binoculars, but mostly just gazing around me. I watch the shifting gleams and glitters of light on the water, the way invisible floating bits and tiny bubbles cast shadows on the smoothly rippled and golden creek bottom, and I listen to the low murmur of water changing direction below the beaver lodge.

Today, as always, I track the bright cirrus and cumulus clouds that float slowly but more often zip quickly across the vivid blue sky. I listen for bird sounds, and though I'm no good at ear-birding, I do know the whir and chirps of the hummingbirds making a racket behind me. I check to see what flowers are in bloom, guessing mostly from the bits of color I can glimpse among the tall grasses and sedges over my head and across the water. Fireweed, I think, a sign of late summer, and two kinds of those umbrellas of tiny white blossoms whose names I have trouble remembering, a bit of yarrow, some kind of yellow daisy-shaped flower. I notice again how the creek bed and the pebbly surface I'm sitting on glitter, too, with the bits of mica fallen away from the mostly granite that makes up the mountains surrounding me. The sun is hot on my skin, but the ground I'm sitting on is cool and so is the slight breeze. It's a perfect day. Here, it often is.

What's different today is in my head: it's been just a few days since I finished reading the manuscript of Ellen's book, the one you've just opened, and so I've got a whole new set of things to look for—and, perhaps more unusually, to imagine without being able to see. I know, for instance, that I should get up close and personal with the dark soil of the bank at my back to look for any sign of the rich underground networks of rootlets and mycorrhizae. Sure enough, I can find lots of tiny, tiny root strands, some of them bright red; a couple of mushroom caps smaller than my smallest fingernail; many minuscule leaves; some grayish webbing and tad of greenish skim. I note how firm this surface is, how even though it looks like packed mud and rises straight up, when I rub my fingers over it almost no bits of soil come loose.

I imagine, as well as I can, the many other microscopic plants and animals that I now know are fulfilling their purposes right in front of me, a much richer array than live just uphill on the valley sides, where dry, rocky forest soils hold their own surprising secrets. I even think for a few minutes about the far tinier molecules of nitrogen and carbon that are stored here—maybe settled in for a long time, time on a geological scale, or maybe just resting for years or decades on their complex paths from wind and rain to the faraway Gulf of Mexico. (Finally, thanks to Ellen, I think I understand *why* nitrogen is so important, not just *that* it is, something I've heard from other scientists I know. Not being a scientist myself, just an interested and curious person who loves being outdoors, I missed or forgot some basics along the way.) Now I know more specifically, and in some way I feel more vividly, how even just the crumbs of rich meadow soil on my palm tie me to much of the rest of the planet, to its past and its future, with the finest of tentacles.

Today, too, I'm scribbling notes, lots of them, things I'm seeing, sensing, remembering from Ellen's book. The sun, light, breeze, colors, glitter. Things moving in the wind, and their sounds. The abandoned but still mucky beaver channels I crossed walking here, the grayed marks of beaver teeth on willow and alder limbs. The one fish I barely saw flit past. The spring's high-water level marked by dried clumps of grass blades, and how snowmelt distributes so many critical elements of this landscape. A leaf floating by, a twig, a mysterious batch of bubbles. What might be blackened pine needles on the creek bottom. Wispy iridescence in some clouds. A question: What made those fine silvery filaments across the creek, the ones that are catching slivers of sunshine and giving them back to the world? How the bottoms of willow leaves flip to silver in a breeze. That there are bumps on the roots around me that collect nitrogen from soil and water and make it into something a plant can use; that willows and alders have all sorts of tricks for surviving with wet roots.

I spot one airplane moving along what is usually a busy flightpath into Denver—but not this year. A yellow bee, a fly, no mosquitoes. In the distance, I can see some of the beetle-killed forest swaths that have become so obvious in these mountains, and there's a light but discernable haze of smoke from wildfires several hundred miles to the west—but today, for a change, I choose not to think about such troubling things. For a little while I venture into the willows until I can see some noisy hummingbirds and discover with amazement that they seem to be sipping willow and alder sap from rows of little square holes in the bark that look like the ones sapsuckers make. I spot a couple of fresh galls on the tips of willow branches that will dry to look just like inch-long pine cones; their incongruity has caught my eye many times, and another scientist friend, a botanist, identified them for me. This is the first time I've seen one when it is still young, soft, and fresh. They look like peach-colored roses. Their texture is that of cactus flowers, a sort of waxy velvet.

For a few minutes I contemplate something else I learned from Ellen a while ago but understand better now—how, beneath the river I see, indeed beneath the floor of this whole small valley, run hidden streams in what scientists call the hyporheic zone, the flows beneath the flow. Finding the gravel and sand paths left by older river routes, avoiding the denser soils from the grassy, sedgy, willowy meadows, these buried streams carry their own insects and fish, their own tiny plants and animals, in the dark for a time, then back into the sunlight. I love knowing this. Like so much else I've just been reading about, the thought of these buried streams reminds me how very alive this place is—and on so many planes and scales, partaking in so many wondrous, intertwining life stories.

As I often do, I think of a writing prompt I learned years ago from illustrated journalist Hannah Hinchman. She asked her students to ask themselves, What all is happening here? By now my catalog for this afternoon in this place would be very long indeed, a list of all the things I've noted here and many more. When I add the breath and blood and electrical impulses and cells moving through my own body, my wandering thoughts and daydreams, the very material but also ephemeral sensations of my skin, my muscles, my hair, all the parts of my own body surrounded by the bodies of all that make up this landscape, this moment, well, words suddenly feel inadequate, though they're my business in life, and my mind falls silent.

I sit very still, breathing in and breathing out, feeling the world in all its intricate parts doing the same around me, welcoming the rays of energy from the sun into my skin and my eyes, into every blade and leaf and drop of water. Knowing more doesn't interfere with these silent, perfect moments. On the contrary, as I've been reminded by my dip into the world of what Ellen knows, it makes them so much richer. For you, dear reader, I predict this book will do the same.

SueEllen Campbell

INTRODUCTION
Hidden Flows Underpin Landscapes

Something hidden. Go and find it. Go and look behind the Ranges—
Something lost behind the Ranges. Lost and waiting for you. Go!
—Rudyard Kipling

What constitutes a landscape? Is a landscape all that you can see from any particular viewing point? Is a landscape a region comprising similar terrain, such as an entire mountain range or flatlands with abundant lakes? Is a landscape an ecosystem defined by interdependent communities of plants and animals? Is a landscape a human-defined entity, such as a national park? Yes. A landscape can be any of these, depending on the context in which the word is used.

The landscape of Rocky Mountain National Park is defined by the presence of the Rocky Mountains. From the Continental Divide, the land falls away to east and west. The continent's protruding spine of rock and the swift drops downward on either side govern the location and movements of everything else. Winter rains and summer snows, the nearly invisible drifts of pollen and dust, the presence of soil, and the ability of plants and animals to move and to thrive all reflect the elevation and steepness of the land. Air flowing east from the Pacific Ocean and northwest from the Gulf of Mexico cools as it rises over the mountains. This cooling condenses water vapor within the air, resulting in winter snowfall and summer thunderstorms. Plants and animals must survive cold air, strong winds, and drifting snow at high elevations, just as other species adapt to limited water and recurring wildfires at lower elevations.

1

Distinctive communities of plants and animals form patterns on the landscape. At the largest scale, the pattern takes the shape of north–south stripes. The alpine community present above tree line at elevations greater than 11,000 feet forms the central stripe atop the Continental Divide. At elevations of 11,000 to 9,000 feet, the subalpine community dominated by spruce-fir forests and lakes forms parallel stripes east and west of the alpine. The montane community of pine forests forms the outermost stripes of the pattern along the lower margins of the national park at elevations of 9,000 to 6,000 feet.

The steep topography within this larger pattern, reflected in the trails that leave hikers gasping for breath as they climb laboriously upward, compresses changes in climate, plants, and animals. Biologist C. Hart Merriam formally described these changes in 1889, noting that climbing upward in elevation on northern Arizona's San Francisco Peaks took him through changes analogous to those observed when traveling thousands of miles northward to the Canadian tundra. Merriam designated life zones that described these changes, and the designations of alpine, subalpine, and montane reflect his ideas.

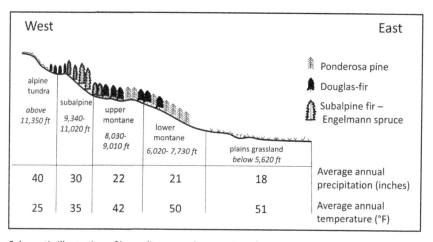

Schematic illustration of how climate and vegetation change with elevation in Rocky Mountain National Park and the adjacent lowlands. In this illustration, the Continental Divide to the west represents the highest elevation.

At progressively finer scales, the landscape resolves into smaller patches of talus slope, icefield, forest, lake, or meadow. Each smaller patch reflects its setting within the alpine, subalpine, or montane, but, like an individual ceramic bead within a diverse and well-crafted necklace, each smaller patch includes distinctive plants and animals that interact in specific ways. The lush magenta hues of flowering Parry's primrose and the sharp *zeet* calls of the water ouzel signal a subalpine stream, while the white and purple chalice flowers of arctic gentian bloom only in the alpine tundra.

The whole is more than the sum of its parts. That expression is well-worn because it succinctly states a widely applicable truth. I am more than a timeline of the events in my life. Rocky Mountain National Park is more than a list of patches or species. So why do we focus on numbers? The national park website lists 276 species of birds, 66 mammals, 11 fish, 2 amphibians, 5 reptiles, and more than 900 species of flowering plants. There is no mention of invertebrates, fungi, microbes, or lichens. This partly reflects the fact that no one knows how many of these species might be present, but it may also reflect the fact that most park visitors would not care. Yet it is these largely invisible species that create the park ecosystems—and views—that we all appreciate. The park's forests could not exist without the interconnected community of fungi present in the soil. Lakes would be sterile water tanks without the microscopic algae that convert sunlight and nutrients into living tissue. Streams would be messy irrigation canals without the microbes, algae, and aquatic insects that build the foundation of a food web from fallen pine needles and shafts of sunlight. Species that remain mostly unseen form a vital part of the hidden flows that underpin the landscape of the national park.

Air and water flow. Air flows violently as 200 mile-per-hour gusts of wind across the Continental Divide, and imperceptibly as oxygen diffusing across the boundaries of a plant cell. Water roars down a mountain stream in a flash flood, and creeps at inches per year through fractures in bedrock. Rock, soil, and living tissues interrupt or assist the flows of air and water, move with them, and—for living organisms—depend on them for life.

Visible and invisible flows link patterns and patches within the park and link the park to the greater world. Wind brings rain and snow, as well as silt and clay to which are attached molecules of carbon, nitrogen, phosphorus, and calcium. Scientists pay a lot of attention to carbon and nitrogen in particular. Carbon gets attention because of its ability to form molecules that can create a greenhouse effect and significantly warm Earth's climate. Humans have greatly increased the carbon content of Earth's atmosphere, largely through burning fossil fuels, and now we seek to understand how ecosystems store carbon and what we might do to enhance that storage. Nitrogen, like phosphorus and calcium, is a nutrient—a substance necessary for living organisms to survive. The movement of nitrogen among the atmosphere, soil, water, and living organisms is one of the natural cycles that people have most significantly altered, with grave consequences for ourselves and other creatures. Human-induced changes in nitrogen show up not only as another potent greenhouse gas, but also as algal blooms in rivers, acidification of lakes, and dead zones in the ocean. Most of the chapters in this book discuss the hidden flows of carbon and nitrogen because of the importance of these elements as critical building blocks of life but also—in different forms and at different concentrations—as pollutants.

Sometimes the flows of air and water leave marks that are visible, but subtle as the trail of silk from a ballooning spiderling. Sometimes the flows create the long-visible scar of unvegetated sediment in the wake of a landslide. Mostly, however, the flows themselves are invisible. No one sees windblown nitrogen accumulating in alpine soils. We see the resulting changes in plant communities on the tundra. The essays in this book focus on the invisible flows within the park and explore how these flows sustain and change the ecosystems of the park at timespans from weathering of bedrock during tens of thousands of years to accumulation of dust on the snowpack during a single spring storm.

Any organism perceives an ecosystem from the confines of its own sensory abilities. We humans mostly see at an intermediate range of the visible spectrum of energy. We cannot see the very small or the very distant, and we cannot see in ultraviolet or infrared. A bee can perceive within the ultraviolet spectrum. A red-tailed hawk can see in the ultraviolet range and detect a mouse from a hundred feet above the ground. A hunting coyote can hear the movements of a vole at the base of the snowpack. A bear can smell another animal's carcass from twenty miles. A woodpecker can detect the vibrations from a tiny beetle burrowing under thick tree bark. The stalk of a flowering plant on the summer tundra can adjust the movement of growth hormones in its cells so that the blossom rotates to track the sun's daily progress across the sky. A tree, connected to surrounding trees by fungal bridges in the soil, can sense when a neighboring tree lacks nutrients and respond to chemical signals of distress by sending nutrients through the fungal bridges.

Any point within the park—or on Earth—can be perceived from multiple dimensions, and these interwoven perceptions create layers of meaning for human understanding of an ecosystem. What we are able to directly perceive is such a minuscule portion of the flows and exchanges within an ecosystem that we can only begin to understand and appreciate that ecosystem if we can somehow become aware of the largely invisible flows that sustain all living organisms.

I am a geologist focused on the study of rivers. I seek to understand how the apparently simple act of water following the pull of gravity down a channel creates the complexities of turbulence, pools and riffles, logjams under which trout take refuge, and flows of carbon and nitrogen dissolved in the water. I can directly measure only a few things in the course of my research. I know how the dimensions of a channel change downstream and how these changes influence the strength of the current and the ability of the flowing water to move sediment and logs. So little, yet this limited insight has great power when combined with the insight gleaned by thousands of other scientists, each of whom delves deeply into an equally narrow portion of the cosmos. My descriptions of the hidden flows that create the landscape of Rocky Mountain National Park derive from the work of this wider community of scientists. None of us has the magical ability to directly perceive all of these hidden flows, but the collective understanding that results from scientific research can make us aware of the existence and vital importance of these flows.

Each of the chapters within this book explores the hidden flows within a specific ecosystem. To make these ecosystems less abstract, I have chosen a particular place in Rocky Mountain National Park to represent each one. Each of these places, however, is a microcosm that represents the macrocosm. The subalpine lake named Loch Vale, for example, is among the most well-studied lakes in the western United States, largely because of decades of work by Jill Baron of the US Geological Survey. Jill and her many colleagues focus on Loch Vale as a starting point for understanding mountain lakes. Each lake, like each person, has some distinctive characteristics, but just as a physician can assume that a healthy person has a body temperature and blood pressure within a limited range, so a lake scientist can assume that a healthy lake has water with dissolved oxygen and nitrogen within a limited range. The hidden flows that I describe in each chapter derive from scientific research conducted in many different geographic locations, but these individual studies build on one another to create a composite picture of how each ecosystem functions.

Ideally, these chapters will be read sequentially, as each builds on information presented in preceding chapters.

Rocky Mountain National Park here also serves as a microcosm for the macrocosm of mountains throughout the temperate latitudes. A hike to the highest peaks of the Appalachians or the European Alps will take you through similar elevational changes. The hidden flows in the soil, the water, and the air described for Rocky Mountain National Park also underpin the landscapes of other mountain ranges.

I have organized individual chapters around specific ecosystems, but this distinction is arbitrary. Although ecologists differentiate montane and subalpine forests, it would be difficult to plant your feet on any spot of ground and declare it the boundary between the two forest types. In reality, a subalpine forest gradually changes to a montane forest at lower elevations, with little fingers of subalpine forest protruding well below the transition along the moist, cool lines of streams. Likewise, although a lake is bounded by a shoreline, streams flow into and out of the lake, carrying dissolved chemicals, sediment, and even logs. Wind respects no boundaries, creating fluxes that reach around the planet. You can never step into the same ecosystem twice, but you can seek insight into the changes large and small that underpin that ecosystem. You can explore and imagine the hidden wonders within the ranges.

1
Montane Forest Hillslope

Midday in mid-July on a south-facing hillslope along Cow Creek in the montane pine forest on the eastern side of Rocky Mountain National Park: there are not many drier places in the park. Snowmelt is finished and the summer afternoon thundershowers have not yet become a reliable event. The morning's dew is long gone, and the temperature has climbed past 70°F. The pine needles underfoot are crisp and dry at the surface. No liquid water is visible, yet water is as critical to every aspect of this ecosystem as in the watery world of a lake or stream.

The understory of the forest is open. Most of the ponderosa and lodgepole pines that dominate this stand of forest grow a few feet apart from their neighbors. Beneath the pines, woody shrubs such as antelope bitterbrush and wax currant concentrate in slight dips along the slope, as do aspen saplings. Kinnikinnick and ground-hugging common juniper form patches of darker green among the hay-colored fallen needles. Green leaves of wild rose hide the dense thorns lining the stem and branches of each plant. Knobs of granite weathered to a pale gray color and covered with lichens protrude above the thick layer of fallen pine needles that cushions footfalls. Walking would be easy but for the numerous downed trees that lie at random angles across the forest floor. Many of the trees still have stout branches protruding from their trunks. Bark remains on some of the fallen trees, but most have developed a silvery gray, smooth exterior over wood that still feels strong despite the

The south-facing montane forest hillslope. Ponderosa pine at left has the reddish-hued bark characteristic of mature trees.

shallow filigrees of beetle tunnels. The absence of bark on the fallen trees is a sign that, although the wood may appear intact, the long, slow process of decay has begun.

This patch of forest has not yet attained old-growth status, which requires at least two hundred years of regrowth after a major disturbance, but sufficient time has elapsed to create multiple signs of longevity. Among these are thick middens of pine scales surrounding the base of some of the trees. Scattered among the pine scales are stripped cores of pine cones that resemble corncobs. These are the leavings of pine squirrels, also known as chickarees, which harvest green ponderosa cones and store them in caches to prevent the release of the pine seeds.

Chickarees tend to be the most easily seen and heard occupants of the forest by day, moving rapidly among the tree trunks and branches in their quest for food and loudly announcing the presence of any human intruder. Chickarees eat the buds, berries, and leaves of plants, but their primary food is the seed of conifers and fleshy fungi. The cones of most conifers mature in the autumn, when the cones open and disperse their seeds. Chickarees interrupt this process, harvesting the seeds while they are still concentrated in the cone. If you have ever seen bits of conifer branches and cones littering the forest floor, as though someone had gone through with an airborne weed-whacker, that is squirrels at work. The squirrels harvest the green cones, clipping them and dropping them to the ground at the rate of one every two to three seconds, and then burying them shallowly at the base of the tree among the pile of cone scales and stripped cones.

A single industrious chickaree can cache thousands of cones each year in a large midden at the base of a tree. Generations of chickarees feeding on selected trees can accumulate a midden that can be up to thirty feet across and more than two feet deep. A midden this big holds moisture and maintains cool temperatures that keep the cached seeds viable for longer periods of time, so some of the seeds that the squirrels miss eating can eventually germinate. The harvest of tree seeds might seem detrimental to the trees, but the seeds stored in a squirrel version of a root cellar may help to regenerate a forest after a severe fire, and the feeding and storage by the squirrels helps to disperse the soil fungi that are critical to conifer survival. In a sense, the squirrels are gardeners, for a white layer of soil fungi is present a few inches below the surface of the midden, and the squirrels eat the fleshy portions of the fungi that protrude above the ground. Through their highly visible activities, the chickarees foster some of the hidden flows that sustain this forest.

The story of hidden flows on this warm, dry hillslope starts with the trees intercepting water, sunlight, and nutrients from air and soil. This is anything but a solo act. The trees depend heavily on microbes in the soil and within their own tree tissues to accomplish these processes, as well as to exchange nutrients with other trees. The story continues with other organisms tapping into the flows hidden within the tree, both while the tree remains alive and after its death.

The Plumbing of Trees

From the perspective of groundwater, rivers are the leftovers. When the ground is saturated, the excess water flows at the surface. There are no leftovers on this south-facing hillslope. The surface may be moist during snowmelt or rainfall, but the trees must seek and hoard water much of the time. How does a tree seek or hoard water? Ecologists and hydrologists, including my colleague Kamini Singha, spy on the secret life of trees. They wire the

Midden created by pine squirrels at the base of a ponderosa pine. Small dark holes in midden are dug by squirrels burying or retrieving cones.

Closer view of stripped cone cores and scales on the midden.

Fungal net a few inches beneath the surface of the midden appears as a white layer.

trunks with electrodes that measure the concentration of chemicals within the tree's tissues and record the trees sucking the nearby soil dry during the course of the day. Other studies encircle the trees with belts that record expansions and contractions of the trunk as the tree absorbs or releases water. Towers that protrude above the forest canopy bristle with instruments to measure carbon dioxide and water vapor, capturing what ecologist Dennis

Baldocchi calls the breathing of the biosphere as trees take up carbon dioxide and release water vapor. The spying reveals impressive plant mechanics.

Trees are tall. A ponderosa pine might grow to be 230 feet tall, a lodgepole 170 feet. This is a long way to lift water from the ground, so the details of plumbing become critical to the survival of a tree. Water enters a tree through the roots. The root cells contain higher concentrations of dissolved material than does the water in the soil, but the cell membranes are selective, allowing water to enter without losing dissolved chemicals. This is sufficient to move the water a short distance up the stem within the xylem that lie within the sapwood under the tree's bark. Xylem act as one set of pipes in the tree's plumbing. The marvelous tensile strength of water molecules allows the tree to move the water up the remaining tens of feet; the tree becomes a living garden hose. Just as water shoots out of a hose because of the pressure created by water flowing in at the other end, so water enters a tree at the roots, moves upward under tension, and exits the leaves at the stomata. Stomata are analogous to millions of tiny mouths on the plant that open and close to regulate the exchange of oxygen, carbon dioxide, and water vapor. Loss of water in the stomata of the tree's leaves and needles creates enough of a pressure difference between the needles and the roots to continue to draw in water at the roots.

A tree must have water because the plant's tissues require it to use the energy in sunlight as part of photosynthesis. Photosynthesis is the chemical magic by which plants exposed to sunlight form ATP, or adenosine triphosphate, a molecule built from carbon, hydrogen, oxygen, nitrogen, and phosphorus. ATP supplies the energy for cellular activities, including changing carbon from carbon dioxide into plant food in the form of a carbohydrate. This is what fundamentally powers the growth of plants.

A plant must exchange gases with the air during photosynthesis, but this exchange has the side effect of losing water. The exchange of gases occurs at the stomata. More than 90 percent of the water that enters the roots of a plant is lost to the air as water vapor when the stomata are open. A person trying to conserve water in the desert does well to keep her mouth closed, as evidenced by the mist that forms when we exhale in temperatures cold enough to condense the water vapor in our breath. Similarly, a plant trying to conserve water does well to keep its stomata closed. But just as a person cannot drink or eat without opening her mouth, so a plant cannot photosynthesize without opening its stomata.

Plants in dry regions develop a variety of strategies to minimize water loss at the stomata. Some plants open the stomata only at night, when the air is less dry. In pines, the leaves have evolved into a needle shape with veins carrying water and sugars in the center, surrounded by photosynthetic cells and then an outer layer of cells that are porous but hard and contain resin. This design helps reduce surface area and minimize water loss, allowing pines to survive long periods of low humidity and to thrive in dry, sandy soils. Other plants keep the stomata closed during the hottest, driest portions of the year.

The stomata must be open sometimes, however, and therefore the tree has to keep taking up water. The center of a tree is virtually dead and remains dry. All the action of moving water and nutrients and growing new plant tissue occurs in the sapwood and cambium that lie just beneath the bark. This living tissue can expand to store water, as revealed in the measurements using belts around tree trunks. Just as a person may need to loosen a belt by a notch after a large meal, belts placed around tree trunks reveal that the trunk swells as the tree takes up water. A tree can adjust how much water it takes up. Trees drink less at night. Flow gauges in small streams indicate that stream flow rises during the night because the trees pull less moisture from the soil. Trees also drink less as the summer and autumn progress. When the weather is dry and the soil also dries, trees become stressed and the stomata close.

In a forest, each tree functions as a water pump, moving moisture from the soil and the underlying cracked bedrock up through the trunk and branches and into the air. Measurements in the forest record the delayed responses to water movements: stream flow rises a few hours after the trees stop releasing water vapor for the day. The source of water used by plants varies with rooting structure and depth. The grasses and other shallow-rooted plants growing across the south-facing hillslope in Rocky Mountain National Park use rainwater and snowmelt coming from just below the soil surface. At greater depths, water molecules can have strong chemical bonds with soil particles or other water molecules within tiny cracks in bedrock, but the pressure created by upward movement of water within a tree can extract even these tightly bound water molecules.

In dry regions, trees use relatively little water. Measured rates of water use by trees range from a frugal rate of 2.6 gallons per day in an oak forest in relatively dry eastern France to a prodigal 312 gallons a day in an overstory tree in the Amazonian rain forest, where there is little need to conserve water.

Lodgepole pines use about five to eleven gallons per day, depending on the age and size of the tree and season of the year. Douglas-fir use approximately five to sixteen gallons per day. These different needs govern which tree species can survive on a dry slope like the one above Cow Creek.

As vitally important as it is for a tree to obtain sufficient water, trees cannot live by water alone. Nitrogen is a critical building block for plant tissue, but it is commonly in short supply and difficult for a plant to obtain directly. The trees of this montane hillslope rely on their hidden partners in the rhizosphere to supply them with nitrogen.

Fungal Nets and Bridges of the Rhizosphere

Astronomers and poets recognize the importance of spheres—spherical planets, spherical orbits, the music of the spheres. Television meteorologists speak of the atmosphere and environmentalists emphasize the importance of the biosphere. But how many of us are aware of the rhizosphere? For plants, life depends on what goes on in the rhizosphere, named in 1904 by the German plant scientist Lorenz Hiltner, from the Greek *rhiza*, for root. This little-known sphere can be simply described as the soil immediately surrounding and influenced by plant roots. That is the only simple thing about the rhizosphere. The nitrogen and phosphorus included within the critical ATP molecules come from the soil, but a plant cannot just suck them up as it can water molecules. Instead, the big trees that impress us so much with their size are intimately dependent on a community of microscopic organisms living within the soil around their roots.

The ability of a plant to grow and stay alive depends on the interactions of its shoots and roots with the surroundings. Between 50 and 80 percent of the carbon that trees consume via photosynthesis may be used to power activities occurring below the surface. Much of this goes into root growth, but a substantial portion also feeds soil microorganisms living in the rhizosphere. This is not just generosity on the part of the plant: the soil microorganisms enhance the plant's ability to absorb nutrients and water, suppress pathogens, and improve the texture and water storage capacity of the soil. Plants can access nitrogen and phosphorus, in particular, mostly because of the ability of soil fungi to extract these nutrients.

Plant roots are the high-density housing within the busy environment of the rhizosphere. Much of the activity of the rhizosphere microorganisms occurs within less than a tenth of an inch from the fine roots of plants. As the tiny rootlets twist and turn their way through the soil, they exude amino acids and carbohydrates that stimulate the growth of bacteria, saprophytes (from the Greek *sapros* for rotten and *phyte* for plant—an organism that lives on dead organic matter), and fungi. These microorganisms in turn produce compounds that repel or stimulate other soil organisms. Just as a terrestrial environment has a food web of plants, herbivores, and carnivores, so the rhizosphere has its own food web. The microflora of bacteria and fungi provide food for grazer herbivores such as mites, nematodes, and springtails, which are in turn eaten by the carnivores of the rhizosphere—centipedes and spiders. All this activity occurs in concentrated areas that are the urban zones of the rhizosphere. And, as in a human city, there is competition for intensely utilized resources, so that the supply of nutrients may limit plant growth.

Each group of organisms within the rhizosphere performs a crucial function. Saprophytes chemically dismantle the remains of dead microbes and roots, decomposing complex organic molecules into basic components. The nutrients released by the saprophytes are captured by mycorrhizal fungi, of which there are thousands of species worldwide. The word mycorrhiza derives from "fungus root." Most land plants depend on these specialized fungi to obtain nutrients. The fungi form a sheath or mantle around plant roots and penetrate a root's outer layer to form a net of nutrient exchange between fungus and host. If you use your hand to scoop through the layer of fallen pine needles into the soil beneath, you may find a moist, patchy layer of white that is the visible manifestation of the mycorrhizal fungi.

The most visible manifestations of the rhizosphere, however, are the fleshy mushrooms that grow on the forest floor and the hidden truffles that are the fruiting bodies of ectomycorrhizal fungus colonies in the soil. The mouthful "ectomycorrhizal" comes from Greek *ektos* for outside, *mykes* for fungus, and *rhiza* for roots. Instead of penetrating the cell walls of the host plants, ectomycorrhizal fungi form a latticework between the cells of a plant's roots. The fungi also form a sheath around the root's surface, with hyphae—the fungus equivalent of tiny, tiny roots—tentacling out a couple of inches into the surrounding soil.

The mycorrhizal fungi send their nearly invisible hyphae far and wide through the soil, seeking out the phosphorus and other essential elements needed by plants. The fungi also urge their host plant to get on with it by producing hormones that promote branching of plant roots. This increases the surface area of the root and enhances both absorption from the soil and contact and exchange between fungus and plant. The plants do their part by shuttling the products of photosynthesis down to their roots, where the mycorrhizal fungi depend on a continuous supply of plant sugars and vitamins. In a sense, the mycorrhizal fungi are subsumed by the plant, functioning as part of the root system and supported by plant energy.

Nitrogen is one of the most vital nutrients that microorganisms in the rhizosphere assist plants in obtaining. Plants cannot use nitrogen by absorbing it directly from the atmosphere. Instead, the nitrogen must be fixed—incorporated with other elements such as oxygen or hydrogen into a molecule that the plant can absorb. So-called nitrogen fixers, the microscopic organisms that create these plant-friendly forms of nitrogen, are like alchemists for plants, creating vital food out of otherwise useless forms of nitrogen. Herbaceous plants such as lupine and trees such as alder have a mutually beneficial relationship with bacteria living within lumps on the plant roots. The microorganisms convert atmospheric nitrogen into a nitrogen compound that is released into the roots of the host plants.

Then there are the free agents among the nitrogen fixers—bacteria living in close association with, but physically separate from, plant roots and mycorrhizae in the soil. Some species of bacteria fix nitrogen in low-oxygen conditions within the soil and within buried wood. Although these free-living bacteria fix less nitrogen than do symbiotic organisms, the steady accumulation of nitrogen associated with their activities makes a difference in plant survival.

So, the saprophytes consume the dead and release food for the net of mycorrhizal fungi just below the soil surface and around the plant roots. The mycorrhizal fungi then trade with the trees—nitrogen and phosphorus to the tree and tree-made sugars and vitamins to the fungi. What we see of all this activity are the white-spotted, scarlet caps of amanita mushrooms or the pale golden bells hanging from the brown stem of the saprophyte known as pine-drops. These colorful manifestations are signposts of the intense underlying activity in the rhizosphere.

The mycorrhizal fungi are busy little biochemical engines, producing organic waste products and rapidly reproducing and dying, all of which attracts other microorganisms that feed in the root zone. Nematodes (roundworms), protozoa (single-celled microscopic animals), and microarthropods such as mites and springtails that graze on fungi and bacteria also release nitrogen in a form available to plants. The microarthropods, in particular, can act as agents of exchange among the high-density housing areas in the rhizosphere by moving fungi, bacteria, and other microbes between sites.

The rhizosphere is a changeable and sometimes challenging environment for its microscopic inhabitants. The pH of soil and soil water can fluctuate substantially as plant roots deplete elements from the soil. Things can get dry as plants suck up water and release it through their leaves. Oxygen can become scarce or disappear when roots and microorganisms use the stored energy of carbohydrates for rapid growth. But there are coping mechanisms. Roots and microorganisms secrete protective gels that can buffer the microbial community from unfavorable conditions. And plants share with each other. Plants connected by a common mycorrhizal fungus can exchange carbon and minerals via a hyphal bridge. If one plant is shaded, carbon may move from the strongly photosynthesizing plant in full light to the shaded plant. The fungal nets and bridges are the transportation infrastructure of the rhizosphere.

Think about that for a minute. The soil is not just mineral grains with a little dead plant material and some plant roots. Vanishingly thin threads of fungal hyphae bridge the physical spaces between masses of plant roots and penetrate the roots themselves. Like utility lines buried beneath city streets that shunt water, electricity, and natural gas across the city and into houses, the fungal nets and bridges keep the plants of a forest functioning through underground connections that, if interrupted, can result in plant stress or death.

Scientists map the fungal nets and bridges using genetics: individual trees are linked if their fungi share the same DNA patterns. These maps reveal that fungal bridges can link more than a dozen individual trees of a single species. The bridges allow mature trees to support the youngsters by transferring water, carbon, and nutrients, and the young trees are inoculated with fungi that help their roots to develop. The largest trees are the most well-connected within this wood-wide web. Canadian forest scientist Kevin Beiler and his colleagues found that a particularly large tree in a stand of Douglas-fir trees in Canada was directly linked to forty-seven other trees through its association with multiple mycorrhizal fungi bridges.

A young sapling linking into this network accesses more abundant resources that help it grow and thrive. The fungi benefit from colonizing multiple trees because their substantial fungal appetite for carbon is unlikely to be met by only smaller, understory trees. Trees linked into the web also have stronger immune systems. Fungal penetration of a plant's roots triggers the plant to produce defensive chemicals, allowing the plant to respond more quickly and effectively to pathogens. Trees within a fungal network can chemically warn one another when insects attack the forest, helping their network-neighbors to withstand the attack. So the rhizosphere is not just the forest equivalent of an underground network of utility lines: it is also the forest's early warning system.

The rhizosphere ages and changes, just as does the overlying forest. Many mycorrhizal fungi strongly prefer specific soil conditions—some prefer buried wood; others prefer exposed mineral soils. As forests mature and the mix of plant species and water and nutrient requirements changes, the species composition of rhizosphere organisms also changes. Disturbances that stress or kill trees, such as wildfire or beetle infestations, ripple into the rhizosphere like waves from a stone tossed into a pond, or reflections seen through the soil, darkly.

Beetle Swarms

So it goes, a delicate dance between resources needed and resources available, in which the microscopic creatures of the rhizosphere and the towering giants of the forest all get by with a little help from their friends. The trees and other plants have their defenses to conserve water, for that is the most common challenge in this place of dryness. Sometimes, however, the defenses are not enough.

A water-stressed tree is vulnerable. Tiny opportunists are continually on the move in the forest, seeking the vulnerable. A tree's living core of sapwood is irresistible to bark beetles and wood-boring insects. Mouth parts adapted for biting and chewing allow both larvae and adults of the diminutive mountain pine beetles, whose Latin genus name *Dendroctonus* means "killer of trees," to chew their way into the hardest wood. If pierced by a mountain pine beetle, a healthy pine may be able to drown the insect in resin. Female pine beetles target drought-stressed trees, and once such a tree is located, the female releases a pheromone that attracts more beetles. A tree under attack can respond with resin and poisonous gases, but sheer force of numbers can

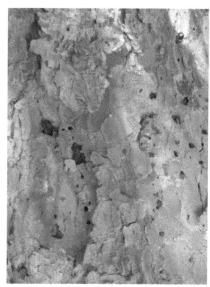

Healthy bark (left) and bark from which the outer layer has fallen away (right), revealing spots where insects and birds have pierced the bark.

overwhelm the tree's defenses as hundreds to thousands of beetles swarm to the site.

Most of a mountain pine beetle's life is spent as a larva feeding in the body of a host tree. A similarly aged group of beetles in a tree complete multiple life stages over the course of one to two years before they all emerge synchronously from the bark to attack new hosts. The time of emergence controls beetle survival. Emerge too soon, and the beetles risk early-season freezes that can kill them. Emerge too late, and they do not have sufficient time to develop before lethally cold autumn or winter temperatures overtake them. Emerge at the optimum time, and the beetles can mount a mass attack that overcomes a pine's defenses, eventually girdling and killing a tree that cannot resist infestation.

The mountain pine beetles move in, laying their eggs under the bark. Larvae hatched from the eggs feed on the living tissue of the tree. What typically kills a tree, however, is the blue stain fungus that the beetles introduce into the tree's plumbing. The fungus decreases and then prevents water transport in the tree, causing a measurable decline in water release from the stomata within a month of infestation. As water transport in the tree's trunk shuts down, stomata close, just as in drought. Closing the stomata results in a decline in photosynthesis because the tree is taking up less carbon dioxide. Eventually,

Tracks of wood-boring beetles revealed once the bark has fallen off a downed aspen log.

the tree dies of carbon starvation, insufficient water movement, or both: the analogy would be a person dying of starvation, dehydration, or both. Once a pine tree dies, the water pump created by the tree shuts down, but the nutrients stored within the tissues of the tree remain.

The death of a tree sets other changes in motion. Pine beetles rarely kill all the trees in a patch of forest. Instead, they target the larger trees, which possess sufficiently thick living tissue to allow beetle larvae to survive a winter nestled inside the tree's tissue. Beetle attack on large trees can create opportunities for the smaller trees to grow more rapidly after a beetle outbreak because of the reduced competition from the big trees. Depending on how many trees die, the rate at which a group of trees releases water from their stomata declines to a greater or lesser extent. This results in more soil moisture for at least the first few months, until rapid growth of understory plants begins to take up more of the newly available water.

Ponderosa and lodgepole pine forests commonly experience water shortages during the middle and late portions of the growing season, but tree death from beetle infestation can produce a temporary bonanza for survivors. Higher water content in the soil reduces the temperature of the soil and the plant litter on the ground surface, as well as reducing daily and seasonal variations in soil moisture. These cooler, consistently moister conditions, along with reduced uptake of nutrients by trees and an increasing rain of nitrogen-rich plant litter from the dead and dying canopy trees, can increase the amount of mineral nitrogen available in the soil within a few months of the beetle outbreak. Like the water bonanza, the nitrogen bonanza lasts until increased rates of understory plant growth deplete some of the nitrogen. A similar surge and then leveling off occurs in carbon content within the soil.

The needles fall from dead pine trees within three to four years of beetle attack. Loss of the needles means that more sunlight, rain, and snow reach the forest floor. Snow, in particular, no longer sits on needle-feathered branches, transforming directly back into atmospheric water vapor without ever falling to the ground. Soil moisture reaches its greatest level earlier in the spring, as the lack of shading changes the timing of snowmelt. Reduced loss of moisture from stomata on the needles of living pines allows soil moisture to remain at higher amounts later into the growing season. The response of surviving trees and understory vegetation to this relative wealth of moisture is critical in determining the carbon and nitrogen balance in the soil during this period, as well as the presence of different species within the rhizosphere.

The dead trees continue to exert an influence, too. Snags that fall to the forest floor add carbon. Carbon from this downed wood, combined with the more readily available carbon in fresh plant litter and compounds exuded from plant roots, reduces the carbon limitations on the microbes of the rhizosphere, stimulating them to greater activity and uptake of nitrogen. This can reduce the nitrogen available for plants, thereby restricting plant growth. But, as snags decompose, the nitrogen taken up by microbes during the decomposition may become available to plants, leading to increases in plant growth. Decay of dead trees does not simply release nutrients to living plants. The situation is more like a tug-of-war between the microbes and the plants, which nonetheless depend on each other for survival.

Recovery of the overall productivity of a forest ecosystem to some steady level may take many decades after an intense wildfire kills all the trees. Recovery can be faster when beetles kill only some of the trees, because the surviving vegetation is able to take advantage of the increased light, nutrients, and water by rapidly increasing growth rates. The devil is in the details: trees use water differently and at different rates in a forest dominated by aspen, for example, than in a forest dominated by lodgepole pine. The winter-bare branches of aspen allow greater amounts of snow to accumulate on the ground than do the needled branches of pines, leading to more water that can run off the surface or infiltrate the soil during snowmelt. Aspen also prefer higher levels of soil moisture and are more prodigal in their use of water, releasing higher amounts back to the air from their stomata than the thrifty pines.

Whether the beetles kill most of the trees in a patch of forest or only a few, the actual death of each tree accelerates the processes of decay that commonly start while the tree is still alive. As with the hidden flows of nutrients in the rhizosphere, the activities of the very small—microscopic creatures and insects—govern the flows of nutrients within the communities that take up residence in the dead tree.

Death and Decay

In a sense, the designation of live versus dead trees is arbitrary. Only about 10 percent of the cells in a live conifer are actually alive. The needles account for 3 percent, the inner bark another 5 percent, and the sapwood the remaining 2 percent. The heartwood that forms the core of the tree is dead. Eventually, however, all of the tree's own cells die, and other living creatures take over its body.

Dead trees go very slowly into their good night in this forest. A lodgepole pine may live for more than two hundred years and a ponderosa for more than five hundred, but that is only a portion of each tree's existence. A dead tree, known as a snag, can remain standing for decades. Many species of birds and mammals rely on snags for food and shelter. Almost half of the birds breeding in Colorado montane forests, for example, use nest holes. Among the cavity-nesters in the montane forest along Cow Creek are mountain chickadees and mountain bluebirds. Their nest holes usually start with woodpeckers.

In the montane elevation zone, three-toed and hairy woodpeckers, in particular, feast on mountain pine beetles. These birds can hear and sense the vibrations caused by beetles tunneling under tree bark. A woodpecker may also use its bill for exploratory tapping to detect hollow or rotten areas on the tree. Their acute hearing might seem to be at risk from the vigorous pounding that occurs once insects are located, but woodpeckers have specially adapted skulls and ear structures to protect both their brain and their hearing from their own hammering. A woodpecker's skull consists of dense but spongy bone encircled by a distinctive array of muscles that, along with the bone, likely help to absorb shock. The membrane covering the opening in the inner ear is also attached to a special bony plate that probably dampens the noise and force of blows with the bill.

The sapwood layer in a ponderosa, which can make up 50 to 75 percent of the tree's volume, decays within a few years after the tree dies, and it is in the sapwood that most of the woodpeckers excavate their nests. A tree cavity housing baby woodpeckers is a lovely sight, but it reflects the combined assault of mechanical fragmentation and chemical alteration of the wood tissue. In the first stage of decay, wood-feeding termites and beetles, followed by woodpeckers, fragment the wood. This increases the surface area available for bacteria and fungi to colonize and chemically alter. The woodpeckers may even give the fungi a ride onto fragmented wood. Foraging woodpeckers are much more likely to have wood-inhabiting yeasts and fungi on their bills than are bird species that do not nest in cavities. The yeast species are those commonly found in soil, and the fungi are common in decaying plants. The yeasts and fungi could be picked up by the woodpeckers while foraging or could represent natural microbial communities found on the bodies of the birds.

Once they reach fragmented wood, bacteria and fungi both extract and add chemicals to the wood, advancing the process of decay. Basidiomycetes fungi (from the Greek words for foundational and fungi), a group that includes most common mushrooms, are responsible for the majority of wood decay. Fungi can decompose wood only if they can penetrate the tree's defenses. The fungi that decompose the heartwood at the center of the tree can enter the plant via diseased roots or via puncture wounds (think woodpeckers) deep enough to penetrate the bark and the sapwood overlying the heartwood. Fungi that decompose the sapwood along the outer portions of a tree trunk have an easier time of it, invading the tree through superficial wounds in the bark and outer sapwood.

From a tree's viewpoint, the cycle is one of adding insult to injury. First the beetles penetrate like little bullets. Then the woodpeckers go after the beetles in a manner more like heavy artillery bombardment, except the heavy artillery contains the equivalent of chemical weapons as the woodpeckers inoculate the tree with fungi capable of decomposing cellulose and destroying the living tissue of the tree.

Decomposition does not occur quickly in the cold temperatures of the Colorado Rockies. Once a live tree or a snag falls, another three hundred to four hundred years may pass before the wood completely decomposes. Intense winds that topple trees and crown fires that kill trees both increase the volume of downed wood on the forest floor. A graph of downed wood volume on the forest floor through time resembles waves on the ocean. A crest in downed wood volume typically forms ten to twenty years after a stand-killing fire, as fire-killed trees gradually fall over. A trough in wood volume follows sixty to eighty-five years after the fire as the supply of standing trees declines. The next crest builds while downed wood gradually accumulates once more over the next two hundred years as trees mature and die.

Like snags, fallen trees are windfalls for a variety of other organisms. Yeasts and fungi move into the wood, as do ants, termites, and other invertebrates. In foraging for those invertebrates, black bears can further fragment the soft, decaying wood. Fallen trees and the roots of stumps provide cover under which rodents nest. In this south-facing montane forest, golden-mantled ground squirrels, southern red-backed voles, least chipmunks, deer mice, and bushy-tailed woodrats all use downed trees as safe havens from which to forage for fungi, fruit, seeds, and insects and other small invertebrates.

Some of the processes associated with dead trees begin while the tree is still alive. Fungi rot the woody material and woodpeckers and other animals excavate the dead parts of the living tree. Conversely, a dead snag or fallen log in an advanced state of decay may contain an abundance of living cells, although these are the living cells of other organisms. Up to 35 percent of the biomass of a dead tree can be composed of fungal cells.

When a tree falls in the forest, greater light, water, and nutrients become available to other organisms. Some of this availability is immediate, as the tree stops shading the forest understory, intercepting rain and snow, and taking up water from the soil. Some of the materials, including vital nitrogen and phosphorus, become available slowly as the tree decays. When a falling snag or tree is uprooted, the roots also lift and mix the soil, making nutrients more available at the ground surface.

Phoenix Trees

A group of standing dead trees left after a beetle infestation might seem like an invitation to wildfire. All those dry trunks appear to be just waiting to

The rootwad of a tree felled by a winter microburst. Note the shallow mass of roots, which peeled up the equally shallow soil like someone peeling back a carpet.

burn steadily, like seasoned logs ready for a campfire, but the situation is not quite that simple. The fire danger does increase during the first four years of a beetle infestation, until the needles fall off the dying trees. After that, ground fires can increase. Surface fuels for a fire accumulate as the needles and then the trees fall. The needles decay rapidly, but downed wood can feed ground fires that burn the accumulated dead plant materials on the forest floor. A ground fire can also reduce the volume of downed wood or largely consume the wood, especially on drier hillslopes. So-called ladder fuels develop as shrubs and seedlings establish and surviving smaller trees grow. These ladder fuels can feed the crown fires that burn through the forest canopy, killing both diseased and healthy trees. Dead and fallen trees create gaps in the forest canopy, however, and reduce the overall density of the canopy. The net effect of all of these changes is that the potential for crown fires drops dramatically within five years of beetle infestation in a forest and then remains low for several decades.

The fact that beetle-killed, standing dead trees can be less likely to host intense wildfires than are living trees is a message that has been slow to spread.

In one striking example from Rocky Mountain National Park, the 2013 Big Meadows Fire on the western side of the park burned about 650 acres before petering out in beetle-killed trees. In contrast, the 2010 Cow Creek Fire on the eastern side of the park, which occurred at about the same time of the year, burned through about 1,500 acres of healthy trees before being contained. Nonetheless, the widespread perception of beetle-killed trees as a fire hazard has led to targeted efforts to cut and remove these trees.

Regardless of drought stress or beetle infestation, lightning can ignite a fire in a seasonally dry site such as the south-facing montane forest, and lightning occurs every summer in Rocky Mountain National Park. In order to survive, both ponderosa and lodgepole pines have adapted to wildfires.

Ponderosa can be long-lived trees, and their bark thickens as they mature. The thick bark acts as an insulator, preserving the underlying living tissue during the heat of a ground fire. The ground fire also kills off the competition for the ponderosas, maintaining the sunny, open understory in which their seedlings thrive. Ground fires may occur every couple of decades in a montane forest, and long-lived ponderosa trees carry the scars from a lifetime of fires in the rings at the base of the tree. Higher severity crown fires kill even the largest and thickest-barked trees, but such fires occur at intervals of hundreds of years and commonly skip about the landscape in a seemingly capricious fashion, killing a large patch of trees on one side of a valley but leaving the trees on the opposing valley side untouched.

In the absence of a crown fire, ponderosas can continue to dominate a forest for centuries, but the spatial variations in fire frequency and intensity leave a forest mosaic with stands of trees of differing species, age, and spatial density. These mixed conifer forests in the montane zone can include ponderosa, lodgepole, and limber pine, Douglas-fir, and aspen, with the abundance and distribution of different tree species reflecting the details of ground and crown fires, beetles, blowdowns, sunlight, and soil moisture. A montane forest forms a collage in which the detailed small patches record the history of disturbances that killed some trees and provided opportunities for others to grow.

Fire must be present for lodgepole pines to persist in a forest. Lodgepole pine seedlings are intolerant of shade and rely on a fire to kill off the competition, even if that competition comes from their parents. In addition, lodgepole pines produce serotinous cones that remain sealed by resin until the heat of a fire melts the resin and releases the seeds. The word serotinous derives from the Latin *serus* for late, presumably because these cones do not open and release their seeds as soon as mature, but rather wait for years until a fire sets the seeds free. Crown fires kill mature lodgepoles but also release the lodgepole seeds. Fifty thousand or more lodgepole seedlings can be crowding each other within a single acre several years after a fire. Competition subsequently thins out the trees as they mature, but lodgepole pines commonly grow in high-density stands—foresters refer to them as "dog-hair" forests—until a crown fire resets the cycle.

Lodgepoles are the phoenixes of the tree world, rising again from the ashes of the fires that kill them. But fire is a dangerous and uncertain ally, and lodgepoles hedge their bets. Some of the cones they produce are not serotinous and can open to release their seeds without trial by fire. Lodgepoles are also adapted to reproduce after a pine beetle infestation or tree topple by wind. The newly abundant sunlight reaching the forest floor can heat the serotinous cones adequately to release the seeds. Seeds previously released from non-serotinous cones may also have lain in wait within the layer of fallen needles and dead plant material on the forest floor, ready to shoot up into seedlings once they receive sufficient sunlight.

In the absence of fire or beetles, trees such as Douglas-fir that germinate well under shade gradually invade the lodgepole stands. In a mixed stand of lodgepole, ponderosa, and Douglas-fir, fires of varying severity can create smaller openings and opportunities for small patches of lodgepole seedlings to establish and persist within the larger patches of other conifers.

In the montane forests of Rocky Mountain National Park, wildfires historically occurred mostly at intervals of thirty to more than a hundred years, rather than frequently, like the surface fires common in the ponderosa forests of Arizona and New Mexico. Fires in the national park's montane forests did not kill all trees over enormous swathes of terrain but instead killed limited stands of trees, creating a forest with patches of trees that also reflect relatively local variations in sunlight and soil moisture. On the north-facing hillslope across the valley, for example, the forest grows more densely in the wetter soil. More Douglas-firs and aspen grow among the pines, and more downed wood is present, even though dead wood decays slightly faster there than on the drier southern slope.

The hidden flows within the soil and the living and dead trees of this forest are like a clock, recording the passage of time as the tree species in the forest and the microbial communities in the soil develop. Wildfire, like tree die-off from beetles, resets the clock. By killing some trees, fires and beetles change the resources available for seeds, seedlings, and other trees, and thus alter the hidden flows within the forest. Violent flows of wind and water can similarly reconfigure the subtler, slower hidden flows.

Too Much Wind, Too Much Water

The four horsemen of the apocalypse for trees in a montane pine forest are beetles, fire, wind, and landslides. Like beetles and fire, wind can be particularly effective in bringing down water-stressed trees. A living tree has impressive flexibility and tensile strength. Both the roots and the stem of a tree can bend. A large crown is more susceptible to the force of wind, but conifers have more tapered and slender crowns than most deciduous trees. Ultimately, however, wood will snap and shatter when subjected to too much force. The most forceful winds occur in the alpine and subalpine zones, where gusts of up to 200 and 90 miles per hour, respectively, have been recorded. Although wind gusts have not been measured in the montane zone, they can certainly be strong enough to snap a pine trunk several feet above the ground, sending enormous splinters of wood shooting into the surroundings. Wind blasts can also uproot the entire tree, ripping up a slab of soil around the shallow roots of a pine. The snapped or toppled pine comes down intact, commonly with needles still green, and the processes of decay take over.

The fourth horseman, landslide, comes with too much water. The great limitation in most environments and at most times in Rocky Mountain National Park is liquid water. Lack of water limits the rate at which repeated freezing and thawing can wedge apart rocks and chemical reactions can change rock into soil. Limited water also constrains the types of plants that can survive and where the plants can grow. And, too little water limits the number and size of rivers and lakes. But, as in most predominantly dry environments, when precipitation comes, it can sometimes come in quantities far too great for the landscape to hold.

The ground is a reservoir for water. Water can be stored in cracks within the bedrock or, in porous rock, within the rock itself. Water can also be stored within the pores present in soil and within the tissues of plants. A montane pine slope in Rocky Mountain National Park is a small reservoir. The limited water available most of the time means that the bedrock is not deeply weathered and cracked and the soil is relatively shallow. The plants, although adapted to store water when it is available, have a finite capacity for taking up water, and the total mass of living plant tissue is small compared to that in a tropical rain forest or a deciduous forest in the wetter eastern United States. When it rains hard and for a prolonged period of time in the montane zone of Rocky Mountain National Park, the small reservoir of the montane hillslope fills up, and something has to give.

The hillslopes can and do shed water that flows downhill in sheets and rivulets, but the sudden abundance of water also lubricates the shallow soil. Masses of soil peel off the underlying bedrock in a landslide that may continue to pick up soil and water as it moves downslope, becoming a thick slurry of rock, soil, and water known as a debris flow. A healthy pine can resist a debris flow moving past its base, although the lower part of the trunk can be scarred by boulders moving with the flow. But a pine cannot resist a landslide or debris flow that effectively pulls the rug out from under it by removing all of the soil in which the tree is rooted.

Somewhere downslope the landslide or debris flow loses momentum and comes to rest in a jumbled mass of sediment and wood. Other forces of change are not likely to modify this mound very much during succeeding years. If the moving material has come to rest across a river course, the river will cut a notch through the sediment without necessarily removing all of it. A beaver may see an opportunity and build a dam on top of the mound. If the landslide or debris flow has come in from the side and blocked only a portion of a river valley, the river may develop a bend and trim back only a little of the mass blocking its path. More than one trail that follows a river in Rocky Mountain National Park has a slight rise and fall as it crosses the mound left by a long-ago debris flow. Gradually, a forest will regrow on the mound. Aspens will likely come first in the wetter areas, perhaps ponderosa and lodgepole in the drier. Within a few decades, these trees may create sufficient shade to deter their own seedlings and favor seedlings of more shade-tolerant species such as Douglas-fir.

The disturbances that kill trees—beetles, fire, blowdowns, and landslides—are explosions within the forest. Large explosions such as beetle epidemics or fire storms can kill extensive swaths of trees; smaller explosions such as blowdowns or landslides kill a few trees or a small patch of forest.

Whatever the extent of the disturbance, other plants and animals move in to tap the hidden flows as the forest starts to regrow.

Death of the Old Guard

Opportunities for such movement are increasing. Overall rates of tree death in old-growth forests in the western United States have increased rapidly in recent decades, with death rates doubling over periods ranging from seventeen to twenty-nine years in different regions. Warming climate is likely contributing to the dying of the old guard. Warming climate can make it easier for mountain pine beetles to survive by limiting the number of extremely cold winter days capable of killing beetle larvae. Warmer temperatures can also correspond to more severe summer thunderstorms with lightning strikes that trigger more intense wildfires. Warming climate can take the form of stronger windstorms that topple trees as air masses from the arctic and the Pacific jostle for position over the Rocky Mountains. Most of all, warmer temperatures lead to limited water supplies and water-stressed trees that are more susceptible to beetles, fire, and blowdown. This is the perfect storm for old-growth conifer forests in the western United States.

As a forest ages, the abundance of large old trees, snags, and fallen wood increases, and their presence provides opportunities for other plants and animals. The height of individual trees within the stand varies, creating a complex vertical structure within the forest canopy. Portions of the forest floor shaded by the big trees attract the saprophyte known as spotted coralroot, which sends up a spike of blossoms delicately shaded in orange, white, and magenta without any greenery. In the more open portions of the forest floor, Oregon grape and heartleaf arnica bloom sunshine-yellow early in the summer. The old trees draw in cavity-nesting birds, small mammals, and mule deer that prefer an open woodland. The dead, fallen trees create ideal denning and resting places for small mammals such as pine martens, which are smaller than the average house cat.

Older trees bear the scars of their long lives in ways both obvious and subtle. The bark of older ponderosas is paler in color and smoother in texture than that of young ponderosas. Rather than the slender, pointed form of youth, the tree is likely to have a sparse or flattened crown. The top of an old tree is more likely to be dead than the tops of younger trees. Broken branches are more common among the aged. Scars from lightning strikes and wildfires, pockets of rot, and burls all record past injuries. The diverse habitat associated with these scars hosts a broader array of other organisms than a young forest. Woodpeckers and other cavity-nesting birds seek out the dead treetops and pockets of rot. Broken branches host different species of lichens than do living branches. The older trees come to resemble their roots in the rhizosphere in the sense of creating islands of biodiversity within the forest.

The forest as a whole becomes a storage zone for organic carbon as the trees age. Carbon is stored in the form of living biomass in the big trees, in the form of dead snags and fallen trees and, most of all, in the soil, rich in partly decomposed plant tissues and the living biomass of the rhizosphere.

The loss of old-growth forests is creating both immediate and sustained changes in the hidden flows from the soil and air into the trees, within the rhizosphere, and among different species of plants and animals. The age and presence of the trees govern carbon dioxide inhaled from the air and water sucked from the soil, the microscopic species present in the rhizosphere and in the bodies of living and dead trees, and the other plants and animals that can thrive in this portion of the forest. Some species will benefit from the decline in older forests, others will suffer. Many of the changes in the hidden flows are noticed only if we watch carefully.

2
Stream Swirls

Streams tie together each of the individual ecosystems within Rocky Mountain National Park and tie the park to adjacent lowlands. Just as the lace on a hiker's boot does not follow a straight course, so a stream does not really flow straight down a mountain. Rather, the water, sediment, dissolved chemicals, and organisms of streams swirl across time and space. To swirl is to move in a twisting or spiraling pattern. Some of these swirls are as obvious as water flowing downstream within the channel. Other swirls—dissolved nutrients moving from the channel into the underlying sediment, or a mayfly newly hatched from a stream and quickly eaten by a songbird—may be less apparent but still represent flows of energy and material within the stream ecosystem or between the stream and the adjacent forest. The story of hidden flows in a stream is mostly the story of all the complications beyond the obvious downstream flow of water.

The water molecules in streams swirl between the atmosphere and the ocean, with the stream as intermediary. Precipitation falling on the catchment of each stream in Rocky Mountain National Park first evaporates from the Atlantic or the Pacific, and some of the water eventually returns to each ocean through stream flow. Shorter swirls form when recently fallen precipitation quickly returns to the atmosphere as evaporation from the soil or a lake, through release from the stomata of a plant, or via sublimation, as snow crystals resting on the ground or on a tree branch transition directly back to water vapor in the dry air of the mountains.

Even if it does not evaporate back into the atmosphere, water in a stream does not just flow down the channel. The sediment in which the channel is formed, and sometimes even the bedrock, is highly permeable, allowing water molecules to also swirl between the stream channel and the subsurface. Some of the swirls are long and slow: groundwater that has gradually accumulated from precipitation falling across the stream's catchment and infiltrating deep into the ground slowly moves into the channel. Stream water can also seep downward to groundwater reservoirs in other portions of the stream.

Other swirls between the stream and the subsurface are short and fast. Each stream is underlain by a subterranean mirror of the surface channel. This hyporheic zone, named from the Greek words for below (*hypo*) and flow (*rheos*), has its own flow patterns and velocities and its own aquatic organisms, from microbes to larval insects and fish. Water enters the hyporheic zone where an obstacle—a large rock, a downed log, the crest of a riffle— creates enough pressure to force some of the surface flow underground. The water flows a few feet or a few hundreds of feet, during a few minutes to a few hours, depending on the stream, and then returns to the surface channel.

The famous adage that you can never step into the same river twice at least partly reflects the fact that the individual molecules of water that start to flow downstream at the headwaters do not necessarily make it out of the stream at its mouth. No one but a hydrologist thinks in these terms, however; most of us conceive of a named stream as a discrete entity rather than a continually changing collection of materials.

The names of streams in the national park reflect the flow of people through the landscape and the characteristics that they noticed and considered important. On the eastern side, the Cache la Poudre River records caching of gunpowder by nineteenth-century French fur trappers. The Big Thompson River may memorialize English fur trapper David Thompson. North St. Vrain Creek takes its name from Ceran St. Vrain, a fur trapper of French descent who built a trading post along the creek near the base of the mountains. These primary channels are fed by tributaries, some of which bear more evocative names, such as Glacier Creek, Icy Brook, Boulder Brook, Beaver Brook, Ouzel Creek, or Cony Creek (cony is an old name for rabbits and pikas). The Colorado River is the primary channel on the western side, orig-

inally named by Spanish Jesuit missionary Eusebio Kino during his explora-
tion and mapping of the river's delta in 1700–1702. Kino apparently named
the river for the reddish color of the water in the lower river, which results
from silt and clay eroded from the iron-bearing rocks along the river's course
downstream from Rocky Mountain National Park.

The tiny, unnamed channels and those large enough to merit a name come
together like strands being braided into a thick rope. Each strand remains
distinct for some distance downstream from its entry point before eventually
merging into an integrated flow, but the integrated flow contains distinctive
signatures from all of its contributors. Even the largest rivers record the pathways by which flowing water laces together the patches of the national park
landscape.

Headwaters

A mountain stream may start at any elevation where snowmelt or rainfall
flowing across the ground collect sufficiently to move soil and form a small
channel, or where a seep or spring reaches the surface with enough flow to
erode the soil and create a channel. The stream may start on alpine tundra,
like Tonahutu Creek on the west side of the park, or in a montane forest near
the lower boundary of the park, like Aspen Brook on the east side. Regardless,
the water makes its way into the larger channels that flow from the park down
the mountain flanks. Streams flowing eastward enter the portion of the Great
Plains drained by the South Platte River. Streams flowing westward enter the
Colorado Plateau and the drainage of the Colorado River.

Designating the start of a stream is more difficult than one might expect.
Flowing water must concentrate sufficiently to allow the water to move sediment and cut at least a tiny channel into Earth's surface. That channel must
persist through time, rather than appearing after each rainstorm and then disappearing during the intervening dry periods, and the channel must continue
downslope, rather than petering out into a smooth hillslope. Channel-like
features seem to be playing hide and seek; they form on a steeply sloping
hillside, continue downstream for a hundred feet or more as a recognizable
feature, and then disappear for another hundred feet or so before reappearing
again farther downslope.

Despite this halting start, at some point down a hillslope there will be
enough water to create a channel that persists through time and down the

slope. The start of this feature is known as the channel head. The persistence of the channel does not mean that it always contains water, however. Erosion can maintain a channel even if the water ceases to flow for part of the year. The stream head marks the point at which water remains in the channel year-round, which may or may not coincide with the channel head.

Both the channel head and the stream head mark a boundary in the landscape. Upslope from the channel head, water, sediment, and dissolved materials move downslope in a diffuse manner. Below the channel head, everything in motion tends to concentrate within the channel or move across the slope toward the channel. Upslope from the stream head, plants and animals living along or using the channel must adapt to at least periodic drying, relying on other sources of water such as groundwater accessed by plant roots. Below the stream head, the continuous presence of water requires different adaptations from plants and aquatic animals, such as special porous tissue in plant roots that helps the roots acquire oxygen in saturated soil. Whether above or below the channel and stream heads, living organisms structure their lives around the presence or absence of water and the amount and timing of the flow of water.

Streams in the Tundra

A stream head in the alpine zone can be obscure. The tundra surface undulates gently, with little hummocks around woody shrubs and depressions where boulders heaved up by freezing and thawing during the Pleistocene Epoch have now slowly subsided back into the soil. These undulations hide the birth-place of a stream—a small spring welling up among the moss and cushion plants. Perhaps only the pink- to magenta-hued flowers of water-loving plants such as queen's or king's crown or rosy paintbrush mark the site of the stream head during summer. The stream flows quietly, without the rush and roar of the larger, steeper channels at lower elevations. Although clumps of tundra plants overhang the banks, the narrow channel is mostly exposed to sunlight and the stream food web starts with attached algae, mosses, and lichens.

Benthic (from the Greek *benthos*, for depth of the sea) organisms are those that live on or in the streambed. Although the phrase "bottom-feeders" has bad connotations in everyday language, the bottom-dwellers literally form the foundation of a stream food web. This foundation starts with biofilms and plants large enough to be readily visible.

A stream coming down from the alpine into the start of the subalpine.

Biofilms are communities of the tiniest—the organisms that individual-
ly are invisible. But the community within a biofilm can be as diverse and
marvelously intertwined as the community of the most majestic old-growth
forest. Within the biofilm live algae attached to the streambed rather than
floating freely in the water. Bacteria, fungi, single cells known as protozo-
ans, and multicellular creatures known as micrometazoans also live in the
biofilm. These organisms live together in a matrix of their own making that
coats cobbles and wood below the waterline. The matrix is the "slime" that
makes streambeds so slippery for people to walk on. For the organisms of the
biofilm, the matrix creates a three-dimensional network that helps the community adhere to surfaces such as cobbles on the streambed and also provides
stability. Like the wide streets and green lawns in a spacious suburban neighborhood, most of the biofilm—about 90 percent—is composed of the support
matrix. Reading the technical descriptions of a biofilm makes the community
sound even more human, as ecologists describe how the matrix limits the
mobility of individual organisms, promoting close proximity, communication between cells, and the "formation of synergistic microconsortia," which
sounds like the prospectus for a start-up company but describes cooperation
among multiple organisms in obtaining food.

In the protective matrix of the biofilm, the different types of organisms create living tissue either via photosynthesis or by taking up dissolved organic matter. It's worth pausing a moment to consider the extraordinarily complex chemical reactions that underlie the simple descriptions of taking up dissolved matter. The biofilm matrix is a chemical gel, created by the organisms of the biofilm, that also acts as an external digestive system for those organisms. Enzymes excreted by the biofilm organisms can continue to exist outside of the organisms. The matrix keeps these enzymes close to the organisms, rather than allowing the enzymes to flow away with the stream water. The continuing presence of these enzymes helps the biofilm organisms collect and ingest dissolved and particulate nutrients from the stream water. The matrix also facilitates recycling by keeping the components of dead cells near the living cells. Basically, the inhabitants of the biofilm excrete a substance that keeps them stable in a fast-moving world, keeps their potential food nearby, and helps them digest that food. Among the food provided by dead cells is DNA, which can be involved in a process called horizontal gene transfer.

Horizontal gene transfer is one of those seemingly outlandish processes that complicate our understanding of evolution and genetics. For generations, we have learned in school how our genes are passed down to us from our ancestors through sexual reproduction. Ostriches create more ostriches and humans create more humans. Except that, as in so many other contexts, we humans tend to forget the tiniest organisms and their influences on our lives. Genetic material can also move "sideways" between one-celled and/or multicelled organisms by mechanisms other than "vertical" transfer from parent to offspring. Scientists now distinguish multiple forms of horizontal gene transfer. In one type of horizontal gene transfer, a cell is genetically altered by the introduction and expression of DNA or RNA from another organism. In another type of transfer, a virus moves bacterial DNA from one bacterium to another. DNA can also be transferred during cell-to-cell contact among some bacteria. As biologist Johann Gogarten describes it, our widely used metaphor of a tree of life should give way to the metaphors of a mosaic to describe the histories reflected in an individual's set of genes and a net to explain the exchange of genetic material among microbes. And it's not just the microbes that are exchanging genetic material sideways: these exchanges account for an estimated 8 percent of human genes. We are more closely related to the slippery biofilm underfoot in a stream than many of us might like to acknowledge.

Biofilms are present in soils and the oceans, as well as in fresh water, and ecologists consider them to be one of the most successful forms of life on Earth, even though most of us aren't even aware of their existence. We are more likely to perceive the other primary photosynthesizers in an alpine stream; the attached aquatic plants large enough to be seen with the naked eye and known as macrophytes (from Greek *macro* for large and *phyte* for plant). In cold, swiftly flowing mountain streams the attached macrophytes are primarily aquatic mosses and related plants known as liverworts.

Feeding on the biofilms, macrophytes, and bits of dead leaves and twigs are the insects that live on and underneath the streambed, in the interstices among the sand, gravel, and cobbles of the hyporheic zone. The larval stages of mayflies, caddisflies, stoneflies, and true flies make up most of these insect communities, but their abundance is limited. Snow and ice cover the small streams of the alpine zone for a significant portion of each year, which keeps the stream inhospitable for many organisms. In lowland, warm-water streams of the eastern United States, a single summer can host two generations of mayflies. In a subalpine forest stream of Rocky Mountain National Park, a single generation can manage to live out its life cycle in one year. In the tundra, aquatic insects need two or more years to get through one generation, just as they do in the Arctic. In the coldest sites, aquatic insects may never reach maturity. They remain present only because of upstream flights of mature insects that lay eggs in the alpine portion of the stream.

When insects emerge from the stream as winged adults, they must run a gamut of predators. Few fish live in the small alpine streams, but spiders build intricate webs among the blades of grass and stems of flowers lining the channel banks, the better to catch the little packets of protein that emerge in summer in the form of delicately winged caddisflies or stoneflies. Birds such as white-crowned sparrows and horned larks eat primarily seeds but will supplement their diet with insects. Similarly, small mammals such as golden-mantled ground squirrels and chipmunks dine on insects, along with a variety of other food.

The alpine streams form ribbons of brighter green stitched across the rocky uplands of the tundra. Flowers are taller and abundant along the streams. The creamy white petals of globeflower surrounding a yellow center bloom early, then give way to magenta and pale yellow paintbrushes, the pink and red hues of queen's and king's crowns, the back-folded lavender petals of shooting star, and clusters of tiny white flowers of bistort. The flowing water underpins the calls of pikas and white-crowned sparrows with a steady rhythm, then quiets as it enters a lake at tree line.

Streams in the Forest

Most of the streams in the national park either flow through the tundra for a relatively short distance before descending below tree line and into the subalpine forests or, like Glacier Creek, start from a subalpine lake. Crossing into the forest creates important shifts in alpine streams. Tall spruce and fir crowd closely to the channel edges, shading much of the channel throughout the day. Although algae remain present, they become less important as a primary source of nutrients. Instead, the big trees shouldering the channel drop their dead tissues—needles, pine cones, leaves, branches, whole trunks—into the stream. This bonanza of nutrients supports the stream food web in forested streams. As the streams collect water from tributaries, however, they grow in volume and speed. A pine needle dropped into swiftly flowing water has

little chance of coming to rest unless something interrupts the flow. The most effective interruption of a stream in the subalpine forest is a logjam that accumulates where a particularly large, fallen tree spans the channel.

A logjam along Glacier Creek exemplifies the complicated structure that can develop around wood in a stream. An old Douglas-fir toppled into the creek during

Glacier Creek in the subalpine zone.

high winds one winter, but a portion of the tree's roots remains buried in the streambank. This anchor makes the fallen tree especially stable, capable of resisting the force of water flowing against it. Branches bristling out from the trunk help retard the force of the flow and catch materials moving downstream. Smaller logs and branches, pine needles, and silt-sized particles suspended in the flow all collect against the upstream side of the fallen tree, building a largely impermeable wall across the stream. Water finds its way around the jam, moving into the porous streambed and then welling up below the jam, and spilling over the top of the jam and over the channel banks during the highest flows. The water following one of these pathways around the jam slows enough to allow even silt and sand and floating pine needles to settle to the streambed, and a mass of fine sediment and bits of dead plants collects in the backwater pool upstream from the jam. This backwater pool is prime habitat for microbes, insects, and fish.

One of my favorite activities is to dip below the water surface in a backwater pool, to a fish-eye view of the stream, courtesy of a waterproof camera mounted on a selfie-stick and set to video mode. In the backwater of the Glacier Creek jam, pine cones, flakes of bark, and needles thickly carpet the margins of the streambed. Pale sand forms an avenue down the center of the pool. Dark-green mosses growing from sunken logs undulate in the gentle current. Bits of decaying leaves float downstream with the erratic flight path of a butterfly. The water surface above is a silvery film, and the light is gentle, diffuse. The dark, irregular wall of the logjam forms a fortress rich in spots to shelter from the current and from predators in the alien world above the water. Immediately downstream from the logjam, swarms of silver air bubbles rush violently toward the streambed and back to the water surface in the steep terrain created by large boulders shouldered up to one another. But here, in the backwater, is a refuge. Just enough current to bring dinner to a waiting fish or aquatic insect, but not so much current that the balance between energy in, from food, and energy out, from exertion, tips the wrong way.

Come winter, ice rims the margins of the pool. Bubbles of air slip beneath the ice and slide along its underside, their shining silvery surfaces reminiscent of drops of mercury. Seen from above, the bubbles form a continually shifting mosaic of white, silver, and black. From below, each bubble acts like a fish-eye camera lens, capturing the curving landscape of snowdrifts and blue sky above the creek. We cannot know how a fish perceives its world, but

Underwater view of the pool upstream from a logjam. Seen from a fish-eye view below the water, the logjam (right) becomes a complex tangle rich in crevices and hiding places. Bryophytes cover the margins of the streambed, and the lighter color at the center of the view is sand.

imagine the shock of being yanked from this cool, protective immersion into the harsh air.

Just as water molecules swirl between atmosphere and surface or between surface and subsurface, so nutrients swirl downstream. The nutrients critical to living organisms—nitrogen and phosphorus—move downstream in fits and starts. They may move swiftly downstream while dissolved in the stream water, then come to a temporary halt as a living organism ingests them or as chemical reactions attach them to a mineral grain or hold them within the tissue of a dead organism. The length of stream along which a molecule of nitrogen or phosphorus makes one complete cycle—from dissolved to living organism then back to dissolved, for example—reflects how much of the available nutrients are being used by stream organisms and the structure of the channel. Shorter cycling lengths equate to more use by stream organisms and to more opportunities for storage, such as behind logjams. A long cycling length indicates that a molecule of the nutrient moves fast and far downstream before a logjam brings it to rest or a larval caddisfly eats it. Thinking about nutrients moving down a stream emphasizes the fact that a natural channel is not an irrigation canal that simply and efficiently passes water downstream, but rather a complex and interconnected ecosystem.

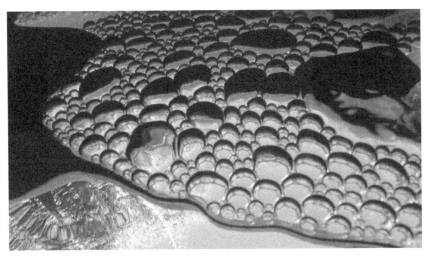

An inverted view of bubbles on the underside of ice rimming Glacier Creek in November. The trees and sky above the creek can be seen through the open water at lower left.

Stream insects do not directly eat an isolated molecule of nitrogen or phosphorus; they eat something containing the nutrients. Insects are like humans in that they eat a wide variety of things, so stream ecologists group the insects based on their mode of obtaining food. Gougers bore into submerged wood to get at the microorganisms living within the wood. Scrapers feed on biofilm, moving across the streambed like miniature sheep and using specialized mouthparts to scrape up the biofilm community and extract nutrients.

Shredders such as stoneflies feed on larger living plants or on dead leaves, needles, and bits of bark greater than a fraction of an inch in size, as well as on the unfortunate microbes associated with the organic matter. Collectors feed on the finest bits of organic matter moving downstream or accumulated on the streambed. Filter-feeders collect moving organic matter from flowing water, either by constructing a tiny net of silk and putting the net into the current like miniature fishermen with gillnets, or by using the fringe of hair along their legs to comb tiny bits of organic matter from the water moving past them. Caddisflies comb the water while partly sheltered in cases of tiny pebbles, or bits of leaf, twig, or needle cemented by silk. These cases are things of beauty and of strength, protecting the vulnerable little insect against the strength of the current and the (to a nymphal insect) boulders of sand grains that come tumbling with the current. Collector-gatherers such as mayflies

Larval insect cases on the side of a boulder in a stream (in Chile).

are more like gleaners, extracting nutrients from organic matter that has at least temporarily come to rest. And, as in any other biotic community, a final set of aquatic insects such as dragonfly larvae live as predators, catching and consuming other insects.

All these insects need to somehow adapt to the swiftly flowing environment of a streambed. Some species of mayfly have evolved flattened bodies, others have streamlined bodies, and still others avoid swift current. Other insects, such as some beetles, never grow beyond a small size and hide in crevices among the sediment grains. Some insects have structures on their feet that allow them to remain stationary: suckers for some midge species and claws and hooks for others; friction pads for some species of mayfly; silk and sticky secretions for black flies; and attachments to rocks for some caddisflies that build cases.

Most aquatic insects gradually metamorphose from nymphs to adults. In some, such as backswimmers, water boatmen, and water striders, the nymphs and adults occupy the same habitat and eat the same food, closely resembling each other except for the lack of wings in the nymphal stage. Other insects such as stoneflies, dragonflies, and mayflies have aquatic nymphs and terrestrial adults with different habitats and food. Mayflies emerge as winged but sexually immature adults that anglers call duns and then, within a few minutes to several days, shed their external skin and transform from a dull-colored dun's body and translucent wings to the shining body and transparent wings of a spinner (in angler-speak). Stream-dwelling beetles remain aquatic as nymphs and adults.

The emergence of the winged adults on a summer morning can be lovely as they move through the mist rising from the stream in the slanting rays of early morning sunlight. I have halted in my work along Glacier Creek to watch as water vapor in the mist churns slowly in the lightest of breezes and the insects flutter upward and downward, their pale bodies like flecks of light in the shadowed valley. They seem to be mayfly fairies dancing at daybreak, creatures of light, air, and foam like the water droplets tossed upward from the swift-flowing stream.

Bottom-dwelling organisms such as mayfly nymphs are not particularly abundant in the headwater streams, but their numbers increase downstream. Part of this increase may reflect the greater diversity of habitats present within the subalpine streams, as well as the greater amount of nutrients entering the stream from the adjacent forest and the surrounding uplands. Rainfall and snowmelt deliver pulses of food to the streams. Nutrients enter the stream dissolved in surface runoff and in subsurface water, and as bits of decaying plant litter carried from the forest floor into the stream.

Abundance fuels abundance. The greater nutrient inputs from the surroundings support greater mass and diversity of aquatic insects. The insects in turn support trout and other fish, as well as ouzels, spiders, bats, and songbirds, among others. Trout get most of the attention, because people like to fish for them. The native species on the east side of Rocky Mountain National Park is the greenback cutthroat trout and, on the west side, its cousin is the Colorado River cutthroat trout. Both species are now rare because of historic overfishing and introduction of other trout species. Starting in the late nineteenth century, people introduced brook trout from the eastern United States, rainbow trout from the Pacific Coast of North America, and brown trout from Europe. Brook trout have been particularly successful and are now the most common trout in the national park. Sharing the subalpine streams with the trout are native western longnose suckers and introduced western white suckers.

Where a waterfall limits the upstream migration of the introduced fish, as at Alberta Falls on Glacier Creek, the cutthroats sometimes continue to dominate the cold, aerated waters of the subalpine stream. These are tough fish, able to survive in the shallowest imaginable liquid water when a small stream freezes and is shaded by deep drifts of snow during the winter months. Cutthroats have been found under boulders and logjams in a stream at water temperatures just above freezing. Juvenile cutthroats can even burrow down a little into the streambed, which then becomes covered in ice frozen to the streambed.

Unquestionably, winter conditions challenge trout survival. The fish can survive winter using behavior and physiology. Cutthroats leave their summer habitats in mid-September, sometimes moving a few miles to find suitable habitat of deep pools with cover provided by downed wood or undercut banks. As ice forms in these sites, adhering to the wood and filling the pool, the trout move again, seeking the ideal winter home of a deep pool covered but not filled by ice, or areas where upwelling groundwater keeps the stream water too warm to freeze.

The formation of ice is not a simple process of water freezing from the surface downward. Because water retains heat longer than air, air cools more rapidly once the sun sets. Water can then lose heat to the air, and on cold, clear nights with temperatures below about 20°F, the water surface can become super-cooled and drop slightly below the freezing point. When the water is flowing turbulently over a cobble or boulder bed, the supercooled surface water can mix throughout the depth of the river. Tiny discs of ice only a fraction of an inch in diameter begin to form in the supercooled water. These discs of ice stick to one another and very quickly the water takes on the appearance of a snowstorm as seen through the headlight beams of a car at night. The globs of ice become larger and more numerous. The rapidly forming ice also adheres to rocks on the streambed, creating anchor ice that can lift cobbles from the bed into the flowing water if the mass of ice becomes large enough.

None of this is good news for trout. The formation of ice that limits habitat and movement is likely the greatest winter threat for trout, rather than cold temperatures or limited food, so the harshest conditions occur where incomplete surface ice cover allows ice to form within the water column. Trout death in winter is particularly likely to occur during the formation and breakup of anchor ice dams as fish are stranded or scraped by suspended ice particles.

Because ideal habitat is limited, large numbers of fish tend to collect in deep pools. For animals that in summer defend territories, coming together in winter marks an important change in behavior. Besides limited habitat, something else may be driving this behavior. Trout also die in winter because of predation. Raptors and animals such as mink continue to hunt actively through the winter, but trout become sluggish and less able to avoid predators. Like giant herds of ungulates seeking protection in numbers, any single fish may be less vulnerable to predators when surrounded by other fish, so the trout relax their territorial instincts and hang together. For trout lucky enough to live in streams that still have beaver, the large ponds that form upstream from a beaver dam provide even better habitat than a logjam and can contain the majority of trout that live within the stream.

Where ice does not completely cover the stream, the fish also hide during the daylight hours. Trout hide under wood in the stream and in the crevices between large boulders on the streambed, with smaller fish positioned in the slower, shallower water closer to the bed. The fish emerge from concealment to feed at night, moving into shallow water with low cover.

Food is in short supply during the winter, so the fish spend most of their time stationary, conserving energy. Trout in cold, high-elevation streams also grow more slowly, reach sexual maturity later, and live longer than fish of the same species in warmer, lower-elevation streams. One advantage of life in the slow lane is that trout populations can persist even if one or more years passes without a new generation of fish growing to maturity.

Cutthroats need cold water to survive the heat of summer, but they do not do well even in large tundra streams because the young fish cannot withstand winter temperatures that are too cold. Once the cutthroats were fish of the forested mountains, period. Now they are largely restricted to the higher elevation, subalpine forests because of competition and displacement from introduced brook trout at lower elevations. The brook trout are highly mobile: they can move more than a mile upstream or downstream on their own, and human anglers have obligingly moved them much, much farther and throughout the historic range of the cutthroats. Both species of trout need deep, abundant pools in which they can rest in the slower water and wait at the edge of the rapid current for insects drifting downstream and into swiftly opening trout lips. Cutthroat and brook trout also need the cover provided by dead wood or undercut banks, as well as places to lay their eggs in the streambed.

Greenback cutthroats, despite their name, are most notable for alternating stripes of pale red and greenish gray along their sides. Brook trout have dark, white spotted backs and reddish bellies, with a distinctive thin white stripe along the outer edge of their lower front fins. This is the stripe that can catch the eye of an eager angler when the fish rest in the shallows.

The trout are silent, their presence notable mostly as a swift shadow darting from beneath an overhanging log when the fish is startled. Ouzels or American dippers, another major predator of benthic insects, are more apparent. John Muir described the ouzel as a "singularly joyous and lovable little fellow." The birds fly swiftly upstream and downstream just above the water surface, creating an aerial mirror of the path and energy of the stream. Sharp cries of *zeet, zeet* pierce the steady sounds of flowing water as an ouzel darts by with rapid beats of its short, sturdy wings. Alighting on a downed log or the slick face of partly submerged boulders, an ouzel seems to delight in the stream. The bird bobs up and down repeatedly before plunging into the swiftly moving whitewater, reappearing a minute later as if by magic, sometimes with insects bristling from its bill.

Ouzels are built for subalpine streams. The chemistry of their blood helps them retain oxygen and remain underwater for minutes, and their dense feathers help them remain warm in the frigid water. Muscles surrounding an ouzel's eyeballs are well developed relative to those of comparable songbirds, so

that the eye can quickly adjust to changes in pressure as the bird moves between water and air. With these adaptations, ouzels either swim or walk along the streambed, gathering insects as they go.

Ouzel on a log beside the stream.

They cannot survive in warm temperatures and seldom stray far from the stream, living in sheltered nests behind a waterfall or beneath an overhanging ledge of rock where the air remains cool and moist. A trout is the underwater embodiment of the streamlined grace and continual motion of flowing water; an ouzel is the aerial embodiment of the stream.

Above the stream hunt the aerial predators of insects, primarily spiders, songbirds, and bats. I like to think of them as dividing the air above and around the stream by elevation and time of day: the spiders on the ground or the over-hanging vegetation, birds such as olive-sided flycatchers and yellow-rumped warblers during the daylight, and bats coming out as the sun sets.

Bats further partition the hunting grounds of the air. The species most likely to be present along subalpine stream corridors are silver-haired bats foraging closest to the ground, little brown bats flying at moderate heights above the ground, and big brown bats flying higher and over wider areas. Each of these species eats various types of insects, although moths form a large part of the diet for silver-haireds and little browns, whereas big browns eat mostly beetles. Silver-haired bats roost in dead trees and leave the Rockies for the winter. Little brown bats roost under bark and rocks, and big brown bats roost in rock crevices. Both species remain in the Rockies year-round, hibernating through the winter. Little brown bats are the stars of the nightly drama above water bodies. As agile as the Harlem Globetrotters performing with a basketball, little brown bats can catch insects from moths to mosqui-toes by striking the insect with a wing tip, catching the insect in the membrane of skin between their hind legs and tail, lifting the insect to the mouth with an upward curve of the tail, and consuming the insect in flight—all during a few moments when the bat appears to tumble briefly in flight. The different species of bats are in turn eaten by everything from frogs and snakes to owls and other raptors, weasels, raccoons, and skunks.

The bits of twig, pine needle, and leaves falling into a stream create the foundation of the stream's food web. Thus, the forest subsidizes the stream by providing food to aquatic organisms. The winged insects emerging from the stream feed spiders, birds, and bats that live in the forest. Thus, the stream subsidizes the forest. Ecologists call these swirls of food and energy between forest and stream "reciprocal subsidies."

Beside the stream, beneath the trees, and sometimes fluttering and danc-ing in the breeze grows a rainbow palette of flowers. Pipsissewa blossom in pink early in the summer, along with purple virgin's bower and the lavender to azure hues of bluebells. Magenta-colored Parry's primrose, white cow parsnip, angelica, and baneberry follow later in the summer, and the baneberry eventually produces vividly tinted scarlet berries. Where the forest canopy lets in sufficient sunlight, the flowering plants create their own dense canopy two to three feet above the ground, completely hiding the irregularities of boulders and fallen logs. Late in the summer, porcini mushrooms, which can reach the size of a loaf of bread, as well as smaller, scarlet-capped fly agaric mushrooms, emerge from the soil. These provide a hint of the subterranean life beneath the forest duff, as do the fruiting bodies of saprophytic pinedrops and spotted coralroot.

Although the most visible mammals along the subalpine portion of Glacier Creek during daylight hours are those that also live in uplands—chickarees, in particular—the stream corridor hosts animals that rely largely on the presence of flowing water. Beavers are chief among these, mainly because of their complete makeover of the surroundings. Beavers are so important in shaping the stream environment where they are present that they get their own chapter in this book.

Other mammals at home in the cold waters of a subalpine stream include the water shrew, which appears in the stream as a silvery sprite, able to swim equally well on top of and beneath the water. The silvery appearance comes from tiny air bubbles trapped in the shrew's short, velvety fur. These bubbles help keep the shrew dry even when it is submerged. Stiff hairs on the shrew's hind feet give the animal the almost magical ability to race across the water surface for short distances without breaking through, like a mammalian water strider. Like the ouzels, shrews feed mainly on aquatic insects, although the shrews can also eat other insects, small fish, and plants. Shrews remain active through the numbingly cold winter, foraging beneath the ice and snow that can help conceal them from predators such as raptors, pine martens, and ermines, although voles and mice make up most of the diet of pine martens and ermines.

Martens and ermines are both long-bodied, slender animals in the weasel family. Ermines are the smallest carnivores in Colorado, and martens are smaller than an average house cat. Although both are most active during the night and the shoulder hours of dawn and dusk, they can sometimes be seen during daylight. I once encountered a pine marten that could not decide how

A weasel (*Mustela* spp.), here in a crevice on a log in the Canadian Arctic.

much of a threat I posed. When I took the marten by surprise as I walked around a bend in the trail along Glacier Creek, the animal jumped up into a tree, looking like an overly large squirrel. The marten then spent more than a minute moving back and forth a few feet above my head while I stood unmoving. Peering at me from the far side of the trunk, darting out onto a branch for a closer look, and quickly jumping between branches, the marten finally leaped to the ground and ran back into the forest with surprising swiftness, leaving me smiling at its beauty and agility.

Martens and ermines are the terror of the voles and mice living in the forest beside a subalpine stream: the western meadow jumping mouse, the long-tailed vole, and the montane vole. The mice are impressively athletic; they can jump straight up to a height several times their size, climb a slender plant stem with swift agility, and swim nearly as well as the water shrews. They fuel their activity by eating a lot of seeds and insects relative to their own small size. The voles, which are strictly herbivores, have stockier bodies and stubby tails compared to mice. The voles create the tunnels visible in matted grass or moss along the stream banks. Their high rates of reproduction allow mice and voles to inadvertently sustain populations of predators in the forest.

Streams in the Woods

As Glacier Creek continues to descend, the surrounding upland forests become drier and more open, with ponderosa and lodgepole pine particularly common. The forest immediately adjacent to the creek remains dense and dominated by tree species that prefer cool, moist conditions—Douglas-fir, spruce, and aspen. Streams throughout the mountains support fingers of subalpine forest poking down into the montane zone among the adjacent, drier upland woodlands, but the subalpine species mix more with deciduous trees such as willow, river birch, and alder that need sunlight and wet soils. In autumn, aspen leaves lie like golden coins on the bed of pools in the stream.

The cooler, moister valley bottoms can still burn, however, just like the adjacent uplands. Much more than a hundred years commonly passes between fires that kill most of the trees within a patch of forest in the subalpine zone. Down in the warmer, drier montane zone, the frequency of stand-killing fires increases, to a century or less. A fire that kills most of the surrounding trees can also significantly alter the stream flowing among those trees. The abrupt increase in sunlight reaching the stream can foster the growth of algae and mosses, shifting the stream food web closer to that of an alpine stream until the shading vegetation along the banks regrows. The flames that consume living trees also destroy the layer of pine duff and the ground plants that help hold the soil in place. The first rains after the fire are likely to carry masses of sediment and charcoal into the stream, filling pools and covering riffles for a period of days to weeks. Bottom-dwelling plants and animals within the stream die in large numbers during these post-fire flushes of sediment, but if enough stream flow is present to remove the extra sediment, the stream community can usually recover in a year or two. Readily mobile organisms like fish may be stressed by the changes, but they survive by moving upstream or downstream to unburned parts of the catchment or finding deep pools that do not completely fill with sediment after the fire.

Flow in the alpine and subalpine portions of a stream is relatively well behaved. The snowmelt of June funnels predictably into the highest flows each year. Although the peak discharge and duration of the snowmelt flood vary from year to year, subalpine streams are less likely to display the exuberance of flash floods in the montane zone.

Flash floods take back the valley bottom adjacent to the channel, spreading across the floodplain and moving the channel itself to a new location on the valley bottom. These floods are fast and unpredictable. Cumulonimbus clouds can gather in the clear blue sky of afternoon within an hour, attain Baroque proportions of height and robustness, pour down water you can prac-tically drink as it falls, and then disappear again with equal swiftness. This is normal. But sometimes the storm does not dissipate so swiftly. The clouds that usually rush overhead somehow stagnate, and the rain falls and falls over one area. Then comes the flood.

Flash floods, especially when they occur in a burned area, have so much energy that the water strips sediment from denuded hillslopes, and the streambed and banks, and sends the sediment churning downstream. Snow-melt floods create a stately dance of the pebbles as the water gradually rises over a period of days, and the sediment on the streambed starts to vibrate in place before individual pebbles and cobbles bounce and roll the length of a riffle or a pool down to a new resting place. The dance of the pebbles becomes more like a stampede when a flash flood mobilizes the entire streambed, flinging boulders up and out of the channel and onto the adjacent floodplain.

The snowmelt flood occurs every year along the entire length of streams in Rocky Mountain National Park, but flash floods are phenomena of the lower elevations. The alpine and subalpine portions of the streams lie at ele-vations too high to receive the concentrated rainfall created when moist air masses moving inland from the Gulf of Mexico release water as they rise over the Rockies. Thunderstorms certainly occur in the alpine and subalpine, but the rising air lacks the abundant moisture necessary to create flash floods. Flash floods are restricted to the montane zone, the foothills, and the plains, where they occur unpredictably at intervals of a few decades in any particular portion of a stream.

In other respects, the montane portion of a stream is similar to the sub-alpine portion. Many of the same species of insects and fish are present, al-though introduced species of trout outcompete the native cutthroats. Brook trout are the most common, but rainbow and brown trout can also be present. Beavers seek the sunnier openings in the riverside conifer forest, where wil-lows, river birch, alders, and cottonwoods thrive. Big conifers falling into or across the channel help form logjams that trap finer sediment and organic material moving down the stream.

Life is a little easier for many organisms in the montane zone, because the winter temperatures are not quite so cold and the stream does not remain snow-covered for as long as in the subalpine zone. Aquatic insects can easily complete their life cycle in one year. Ouzels remain present along the stream in the upper range of the montane zone, but different mammals appear along the banks.

Montane voles are active day and night, year-round, repeatedly emerging from their globular grass nests to eat grasses, forbs, sedges, and fungi growing in the moist soil along the stream. The voles leave signatures of their activities in the form of well-developed runways through the thick grass and sinuous mounds of soil. The soil mounds, which resemble fat mineral snakes, form in the network of tunnels that voles construct at the base of the snowpack during winter. If you pick up a section of these soil-snakes, it sits on solid ground. When the surface of the snowpack warms in spring, meltwater filters down and flows along the ground surface, carrying with it some of the silt and sand displaced by the burrowing of the voles. The meltwater concentrates in the snow tunnels excavated by the voles, gradually filling these tunnels with sediment. When the snowpack fully melts, the sediment-filled tunnels are clearly visible, but they gradually smear out into diffuse piles of sediment as the summer progresses.

Masked shrews, which are also present in the subalpine, live in moist meadows and willow thickets along the stream, where they eat mostly in-

A line of sediment that has filled subnivean rodent tunnels during snowmelt.

sects. Southern red-backed voles can also be present in willow thickets along the stream. These voles are particularly fond of eating fungi, some of which they find in the middens of chickarees. Where the voles are silent (at least to human ears) and surreptitious, the chickarees are vocal and boisterous. Their repetitive, aggressive chattering reflects the strongly territorial nature of the squirrels, perhaps because they work so hard to maintain a food supply to fuel their year-round activity.

Chickarees are less common in stands of ponderosa pine occupied by Abert's squirrels, which are readily distinguished by the tassels of fur that make their ears look pointed. Abert's might as well be called ponderosa squirrels, for they depend on the trees for both nesting sites and food. Inner bark, buds, twigs, seeds, and young cones of ponderosa are all palatable to the squirrels, but, like chickarees, Abert's also eat fungi and help disperse the spores of mycorrhizal fungi through their feces. Also like the chickarees, the activity of an Abert's far overhead is revealed by a pile of severed needle clusters and bare twigs at the base of a tree, but the Abert's do not cache food.

Sharing the canopy with the squirrels are birds by day and bats by night. Hoary bats particularly favor ponderosa groves, where they roost in the trees and emerge after dark to feed, primarily on moths, and to be preyed on by owls and hawks. Moths can see exceptionally well at night, and bats can of course find the moths using echolocation. The St. Lawrence tiger moth is among the species favoring riparian areas. This chocolate brown and orange moth with cream-colored spots ranges widely across higher latitudes and altitudes in North America. Moth larvae feed on the willows, alders, and birches that also favor riverside habitat.

The banks of the stream are bright with flowers throughout the summer. The intricate shapes of lavender-hued fairy slipper orchids, the covert white bells of twisted stalk, pale blue and white Colorado columbine, purple monkshood, and white wood nymph flowers appear early in the summer, followed later by broadleaved bluebells and tall stalks filled with the flowers of white bog orchid. In shaded areas rich in decaying wood just back from the banks, the reddish-brown, drooping bulbs of pinedrops and tiny, red-splotched white petals of spotted coralroot appear by midsummer. Later in the season, clusters of the startlingly scarlet berries of baneberry emerge among the large green leaves of the plant. The montane stream feeds all of these plants as the stream water seeps into the banks and creates moist soil.

The opening within the forest created by the stream allows sunlight to reach the forest floor.

Swirls

Glacier Creek flows down through these diverse communities of plants and animals into a progressively warmer and drier climate before joining the Big Thompson River and continuing beyond the boundaries of the national park. The creek retains signatures of this journey. Dissolved chemical compounds of nitrogen, carbon, and phosphorus reflect inputs from rain, snow, and windblown dust, as well as the subterranean pathways the water followed before joining the creek. The amount of flow throughout the summer largely reflects the amount of snow that fell during the preceding winter and how rapidly that snow melted and moved downslope into the stream. The abundance of sand, silt, and clay, as well as cobbles and boulders, records the history of wildfire and debris flow, landslide and rockfall, and simple grain-by-grain movements downslope from hillslopes into the stream. Bits of wood and pine needles integrate the litter dropped by plants along the stream's path and the ability of the stream to trap or transport that litter downstream. Each of these signatures is as distinctive as a fingerprint, and scientific detectives can trace the paths of water, sediment, and organic matter entering the creek and joining the continual downstream flows of material. Aquatic organisms also shed their tissues in the stream, leaving signatures that can be read in environmental DNA.

Environmental DNA, known as eDNA, can be collected from any material—soil, ocean water, snow, or air. The DNA itself is the same as that collected from living or dead organisms, but eDNA results from the cells continually shed by organisms: specifically, the cells shed in mucus, feces, skin, hair, and other material from animals, as well as bits of leaf, flower, seed, or stem from plants. These bits of effluvia can be used in DNA sequencing, in which DNA in a sample is compared to mapped sequences in DNA libraries to identify organisms present in the environment. Among the many extraordinary benefits of eDNA is the ability to detect the presence of a species without having to capture an organism and remove tissue for DNA analysis and identification. This could be the proverbial game-changer for mapping the presence of organisms that are rare and potentially dispersed over large areas, are very difficult to find and temporarily capture, or could be seriously stressed by

The South Fork Poudre River in the montane zone.

physical handling associated with capture. We all leave traces of ourselves in ways that we do not imagine.

Glacier Creek and all the mountain streams of the national park, following the inexorable dictates of gravity, lace together the alpine, subalpine, and montane communities, each of which leaves its mark on the streams as the water continues down into the plains and ultimately to the ocean. This stream-lacing involves abundant swirls. The lace on a hiking boot swirls from side to side of the boot, up and down as the lace repeatedly crosses through the fabric of the upper boot, and from the boot-top to the toe and back again. So a stream swirls from side to side through the mountains, lacing together the land and stream environments: water and sediment move into and out of the stream, the bodies of plants and animals create reciprocal subsidies, and nutrients swirl between dissolved and particulate forms and the tissue of living organisms. The stream swirls upward and downward as water and dissolved chemicals move among the air, the stream channel, the underlying hyporheic zone, and the deeper groundwater. And the stream swirls along its length as water, sediment, and dissolved chemicals move gradually downstream and some organisms move upstream. Just as a hiker is hard put to move well with an unlaced boot, so the ecosystems of Rocky Mountain National Park are hampered if the hidden flows that underlie the swirls of the streams come undone.

3
Beaver Meadow

The streams of Rocky Mountain National Park flow steeply downward from the heights of the Continental Divide. They tumble down stepped channels and plunge over the lips of waterfalls, their water white with the foam of air they gulp in their tumultuous descent. The only pauses come where a glacier once abraded the valley floor, gouging a broad-bottomed trough into the bedrock. In these gentler sections of valley, the streams metaphorically pause and take a breather. They meander across the valley bottom, soak into the sediments and flow a short distance underground, and partly lose themselves wandering among smaller channels and into ponds. Some of these streams become bewitched by the magic of beavers and almost cease to resemble streams.

The magic of beavers might seem obvious. Dams bristling with sticks that the beavers have chewed into sharply pointed ends punctuate the stream and surrounding floodplain, creating ponds large and small. The steep-sided canals one beaver-body-wide that the rodents dig to help them get safely around the valley bottom are a little more hidden, precipitating a fall by unwary humans crossing the beaver's territory. But the beavers also change the hidden flows of the river and floodplain, creating an environment that stores water, sediment, and nutrients for longer periods of time than portions of the river without beavers, and creating habitat for so many other plants and animals that the flows of food-energy between members of the river community are fundamentally different where beavers are present.

Beavers through Time

When the cold, dry climate of the late Pleistocene began to warm about 16,000 years ago, the glaciers that occupied the upper valleys on both sides of the Continental Divide in Rocky Mountain National Park started to lose ground. Year by year, the ice flowing down from the head of each valley did not extend quite as far down-valley, so that the front of the glacial ice effectively retreated through time. Terrain that had been covered by massive ice for thousands of years now received the meltwater and sediment released by the retreating glaciers, and pioneering plants began to colonize the new land. Grasses, lichens, and other nonwoody plants came first, stabilizing the newly deposited sediment and starting the formation of soil.

The tongue of glacial ice within each valley retreated in fits and starts. Stable at one position for decades, the ice released sediment that formed mounds along the sides of the ice, known as lateral moraines (or an end moraine in the form of a ridge across the valley), before retreating hundreds of yards upstream and temporarily stabilizing once more. House-sized chunks of ice remained as the ice front retreated. Sediment deposited around and on top of the ice chunks insulated them from the sun's warmth and slowed the rate of melt. When the ice finally melted, overlying material collapsed into the new cavity, creating small lakes that gradually filled with sediment. Along with the water ponded above end moraines, the ice-melt cavities created wetlands in which decaying plants enriched the soil. Gradually, woody shrubs and trees colonized the valley bottoms, providing the food to support beavers.

Scientists who study ancient ecosystems infer that beavers colonized landscapes around the glacial margins within four thousand years of the ice retreat. In Rocky Mountain National Park, that means that beavers have likely occupied valley bottoms such as Moraine Park or Horseshoe Park for more than ten thousand years. These wide, gentle portions of valley bottom tend to have wet meadows and willow thickets rather than the conifer forests of steeper, well-drained valleys. Nineteenth-century fur trappers referred to such areas as parks or holes, as in Brown's Park on the Green River in western Colorado or Jackson Hole in Wyoming.

The abundant habitat and food of a wide, wet valley bottom makes it ideal beaver country, and the animals prospered in these sites for thousands of years. Then, two hundred years ago, men seeking furs to support the fashion habits of other men entered the Rocky Mountains with traps and guns. These men did not seek to kill a few beavers: they largely eradicated the animals. In 1842, John Charles Frémont traveled through the region that would become Rocky Mountain National Park. He described the ruins of the beavers' world—abandoned lodges and dams falling into disrepair. Only occasionally did he sight a survivor.

Fortunately for the beavers, fashions changed, and the survivors recolonized the landscape. Beaver populations in the national park gradually rebounded until around the 1960s, when a new threat started to limit their survival. Elk, native to the park but hunted to extinction early in the twentieth century, were reintroduced into the park from Yellowstone in 1913. Predators such as wolves had also been hunted to extinction in Rocky Mountain National Park and the surrounding region, but the predators were not reintroduced. In the absence of hunting pressure, elk populations exploded. Moose were introduced just north of the national park in 1979 and, being strong, long-legged wanderers, readily made their way into the park and prospered. The intensive grazing of riparian plants by elk and moose so severely limited food for beavers that after the 1960s beavers once again largely disappeared from the national park.

They did not disappear completely, however. A healthy population persists in Wild Basin along North St. Vrain Creek, and smaller colonies are present elsewhere on the east side of the park at Lily Lake, Glacier Creek, Hidden Valley, Horseshoe Park, and Cow Creek. These are the source populations from which new colonists go forth each year. The mated pair in a beaver colony has between two and four kits each year. When they reach two years of age, the young beavers strike out on their own. This is a dangerous time for them, and many die. But the fortunate or skilled few go on to establish a new colony and work their beaver magic on a new section of stream. In Rocky Mountain National Park, any site chosen by a two-year-old beaver has likely been occupied by other beavers in the past.

Convoluted Plumbing

If one of the parks along a stream in the national park has been abandoned by beavers for more than a decade, the stream flows in a single channel that

A beaver dam on Hague Creek in Rocky Mountain National Park.

typically has cut down below the surface of the adjacent meadow. The portion of Beaver Brook known as Upper Beaver Meadows—now uninhabited by beavers—provides an example. A fringe of shrubby willows, heavily browsed by elk, lines the channel. Although the meadow has sufficient water to support a new growth of emerald-green grasses each year in June, the soil is drier than that along a stream dammed by beavers.

If the remaining willows are sufficiently large and abundant, their stems can provide the building material for a new beaver dam. First the uprights—four- or five-foot lengths of willow stems—are anchored into the streambed. Then the beaver fills the gaps between the uprights with smaller wood pieces, mud, cobbles, branches, and leaves—any available materials that will stop the flow of water through the dam and create a pond in which the beaver can build a lodge. Where steep stream banks occur, the beaver may dig a den into the bank and use the dam to create sufficiently high water levels to keep the den entrance submerged. Lodge or den, the objective of the dam is to create water deep enough to allow the beaver to move about more safely and keep the entrance to the lodge submerged to deter predators.

The active beaver meadow along North St. Vrain Creek in Rocky Mountain National Park. This view, in August, shows abundant water ponded by beaver dams on the floodplain.

As dam and den take shape, the beaver establishes a network of trails and mounds of grass and mud marked with its own scent glands to delineate its territory. If all goes well, the new colonist finds a mate, and a family of beavers occupies the site. The growing colony excavates a network of canals across the valley bottom that allows them to move about safely. Additional dams are built on the main channel or on tributaries coming into the valley bottom, or even on valley-side seeps or springs. Any place that has enough water flow to eventually support a pond is suitable. As the work of the beavers continues, the valley bottom becomes a beaver meadow—a complex of dams, ponds, smaller channels, and dense thickets of willows that spread across the entire valley rather than being limited to a line along the main channel. At present, North St. Vrain Creek in Wild Basin hosts the only extensive, healthy beaver meadow left anywhere in Rocky Mountain National Park.

Each beaver dam is a punctuation mark in the flow of North St. Vrain Creek. As long as water keeps flowing, the energy of movement carries along sand, silt, and clay, as well as leaves, pine needles, twigs, and dissolved carbon and nitrogen. Where the water slows in a pond, all of this material settles onto the floor of the pond. Sedges, rushes, and water lilies take root in the rich muck on the pond floor and the plants act like magnets, drawing in insects and birds.

During snowmelt and summer thunderstorms, increased flow into the pond overtops the banks and spreads across the valley bottom. Some of this water filters into the floodplain, filling minuscule spaces between sediment grains to saturation. Some of the water concentrates in the beaver canals, flowing more swiftly than the water outside the canal and creating enough force to erode the sides of the canal and turn it into a secondary stream channel. Flow from the main stream also filters into the stream banks and streambed, entering the hyporheic zone or the underlying groundwater.

Flow in a stream reflects a balance between energy driving water downstream and forces resisting the flow of the water. The driving energy results from the mass of water and the slope down which the water flows. Forces resisting the flow of water result from the characteristics of the stream banks and bed: big boulders resist the flowing water better than sand grains, and sandy stream banks knitted together by willow roots resist the energy of the stream flow better than unvegetated banks.

Hyporheic flow also reflects a balance between the driving energy of water and the resistance of buried sediments. Sand and gravel have larger empty spaces between individual grains, and subsurface water moves easily through these materials. Silt and clay stick together and pack so tightly that there is little space through which water can move. Consequently, hyporheic flow paths largely follow patterns dictated by sediment size. If an underground layer of silt and clay parallels the surface stream, there will be minimal flow in the subsurface beside the stream. Conversely, if a band of gravel is present below the surface far from the stream, this gravel is likely to carry substantial hyporheic flow that originated in the distant surface stream and will return to that surface stream at some point downstream.

The complexities of subsurface sediment and hyporheic flow patterns result from the history of the creek. Stream channels naturally shift back and forth across a valley bottom through time. If the stream meanders, faster velocity at the outside of each bend erodes the bank even while slow velocity at the inside of the bend allows sediment to deposit and form a point bar. With time, the channel migrates along the outer bend and leaves a trail of point bar sediment at the inner bend until the whole bend is cut off during a large flood and a new bend starts to form elsewhere. The former bend, chopped off from the main channel, can form a pond that gradually fills with sediment and bits of dead plants.

Beavers accelerate the movement of a channel. Although a beaver dam slows flow upstream, water spilling over the banks during high flows can erode the banks. Water concentrating in the beaver canals can form new channels. As a pond fills with sediment and bits of dead plants, the beavers build a new dam upstream or downstream or on a secondary channel. Silt and clay accumulate in the abandoned pond, but the newly formed surface channels elsewhere on the floodplain are more likely to have a bed of sand and gravel, until these channels are also dammed and fill with silt and clay. All these processes result in an extremely complex, three-dimensional mosaic of surface and subsurface flow paths that significantly challenge the ability of a hydrologist to map and understand them.

Simply measuring the flow in the surface channels is not enough. Such measurements show that more water comes in to the beaver meadow during peak snowmelt than flows out at the downstream end. The water is stored somewhere through the summer and then gradually drains late in the season, with more water flowing out of the beaver meadow than is entering upstream during autumn. Some of the water is stored in ponds and the hyporheic zone, but some is also lost. Water evaporates from the ponds and channels and escapes from the stomata of the plants. Some of the subsurface water may also keep going downward, entering the groundwater through fractures in the bedrock underlying the valley and ultimately remaining there or flowing elsewhere in the landscape.

Understanding the pathways of water that remains in storage for a month or more is also difficult. An off-channel pond with no surface connection to a channel can follow rising and declining stream flow in slow motion: the water level of the pond rises a week after flow peaks in the stream and then declines in an equally lagged fashion. Faced with this complexity, hydrologists respond with an array of tools. Instruments deployed across the valley bottom can be used to record times of rise and fall of surface and subsurface flow in channels, ponds, and the hyporheic layer. Tracers in the form of salt or colored dye can be added to the water to follow specific pathways. Energy waves—radar, electrical waves, weak seismic waves—sent into the ground can be used to create maps of underground grain size and moisture content. Water temperature can be measured along streambeds: cooler temperatures indicate sites where hyporheic water is returning to the surface.

All these approaches are useful, but like a pair of binoculars that will not quite focus correctly, they still provide only a fuzzy image of the mysterious paths by which water moves through a beaver meadow. What does emerge more clearly is that, although water weaves paths between surface and subsurface, groundwater and atmosphere, the material that the water carries remains in the beaver meadow.

The Sediment Stops Here

Water flowing downstream along most portions of the channels in Rocky Mountain National Park has more than enough energy to transport sand, silt, and clay. The primary limitation to this transport is the supply of this smaller sediment. Most of the length of streams in the national park is floored with cobbles and boulders that require more energy to move. These coarser sediments move each year during snowmelt runoff, especially during years of unusually high flow. When the bigger grains are dislodged from the streambed, they can expose the underlying sand and silt and allow that smaller sediment to also move. Snowmelt and rainfall running over the ground surface also carry smaller-sized sediment into the streams. In the absence of a lake or a beaver pond, most of this sediment just keeps going until high water overtops the banks and drops sediment in the slower, shallow flow of the floodplain or until the stream flows into the flatlands beyond the mountains.

Any small sediment in transport that enters a beaver meadow is likely to remain there for at least three reasons. First, the beaver dams and the ponded water upstream from each dam create a barrier to sediment movement. Even if a beaver dam breaches and the pond drains, sediment eroded from the base of the pond is unlikely to go far downstream before entering another beaver pond. Second, the presence of multiple channels that branch and rejoin in a beaver meadow reduces the ability of each channel to transport sediment. If fifty cubic feet of water per second (cfs) flows into a beaver meadow in a single channel, that flow expends some energy overcoming frictional resistance along the bed and banks of the stream. If that fifty cfs is then split into three channels within the beaver meadow, each channel may carry only ten to twenty cfs. A greater proportion of the water in a smaller channel is exposed

The eroded and exposed end of an old beaver dam, showing fine sediment and chewed sticks. This dam is along North St. Vrain Creek in Rocky Mountain National Park.

to frictional resistance from the channel boundaries because the ratio of surface area (bed plus banks) to water volume is greater in a smaller channel. In other words, an equal volume of flow expends more of its energy simply continuing downstream in a series of smaller channels than in a single, larger channel. This means that less energy is available to carry sediment.

Finally, the vegetation growing across the floodplain in a beaver meadow limits downstream sediment movement. Willows, alder, and river birch thrive in beaver meadows. These plants tend to grow in dense thickets. When water overtops the channel banks during a flood, the closely growing stems of woody plants, along with rushes and sedges, create so much frictional resistance that the flow slows down and drops any sediment it is carrying. Floodplain plants can respond quickly to this deposition, growing rapidly and effectively anchoring the newly deposited sediment with their roots. Through all these mechanisms, sediment entering a beaver meadow is likely to stay there as long as the beavers actively maintain their dams and keep the soil wet enough to support dense willow thickets.

The beaver meadow also retains bits of leaves, twigs, and dead plants. This organic matter rich in carbon, nitrogen, and other nutrients then slowly de-

Willows in a beaver meadow along a secondary channel of Glacier Creek in Rocky Mountain National Park. In this early June view, the willows are just starting to leaf out.

cays to create fertile soil. Anything attached to mineral sediment—carbon, nitrogen, phosphorus—also remains in the beaver meadow with the sediment. Some portion of these nutrients becomes available to plants. A substantial portion of the nutrients is also stored, as long as the sediment remains saturated, because fewer species of microbes that ingest these nutrients can live in sediment without oxygen. Microbial communities in the floodplain soil and in the hyporheic sediment largely govern the fate of nutrients within the beaver meadow.

iNvisible

Microbes are the invisible masters of the North St. Vrain beaver meadow. To focus on just one example, their activities govern the distribution of critically important nitrogen (abbreviated as N in the periodic table). Most organisms need nitrogen to survive. When nitrogen is limited, plant growth is restricted and ecological productivity declines. But excess nitrogen is equally problematic, causing algal blooms that deplete the oxygen dissolved within water and leading to "dead zones" in which most organisms cannot survive.

Most nitrogen circulating through ecosystems comes from atmospheric nitrogen gas (N_2), but the gas is inaccessible to plants and animals—you can't just suck in a big mouthful of N_2 to get the nitrogen your body needs. Instead, plants and animals rely on bacteria to chemically alter the nitrogen into a usable form. Two groups of bacteria start the alterations; bacteria that live within the roots of legume plants, and soil bacteria that convert nitrogen gas to ammonium (NH_4^+). Other species of soil bacteria then convert ammonium to nitrite (NO_2^-). Nitrite is unstable and quickly latches onto an additional oxygen atom to become nitrate (NO_3^-). In this form nitrogen is a nutrient readily available to plants. Soil bacteria are like workers on an assembly line, passing the nitrogen along and altering it at each step, except that the process is not linear: nitrogen is constantly being exchanged among the atmosphere, soil, water, and living organisms.

Nitrogen enters the beaver meadow from literally all directions. Some comes directly from the atmosphere and some comes in with rain and snow. The pathways of the rain and snow matter: precipitation falling onto a forest canopy can dissolve nitrogen from the tree leaves before continuing into a stream. Dissolved and particulate nitrogen flow into the beaver meadow from upstream portions of the river network. Leaves and twigs that contain nitrogen within their tissue fall into the stream from the overhanging trees and shrubs or fall to the ground and are then carried into the stream during snowmelt runoff. Nitrogen can also enter the beaver meadow from the subsurface via groundwater.

Nitrogen leaves the beaver meadow by equally diverse pathways. Some nitrogen moves downstream in dissolved or particulate form within the stream water. Some returns to the atmosphere as ammonium released from the stream water, or within the tissue of a mayfly or dragonfly that emerges from the water to become a winged adult. Some of the nitrogen goes back into the atmosphere in the form of N_2 produced by yet another set of bacteria that live in environments with no oxygen gas. The remainder of the nitrogen is either consumed by aquatic plants and animals and stored within living or dead tissue or attaches to silt and clay particles and enters the streambed and floodplain.

Nitrogen consumption by living organisms or storage within sediment depends strongly on the activities of the soil bacteria. These bacteria are most effective at consuming nitrates within the soil beneath riverside plants, within the hyporheic water, and within the sediment of wetlands such as beaver ponds. Because beavers create environments favorable to bacterial consumption of nitrates, beaver meadows function as nitrate sinks that limit the amount of this potentially damaging chemical that continues downstream.

The balance among nitrogen entering, leaving, and stored within the beaver meadow changes over hours to days and years to millennia, and beavers are the regulators of this balance. A beaver dam creates a difference in pressure from the stream to the subsurface that enhances the downwelling of water from the stream into the hyporheic zone, where microbes can remove nitrate from the water. The dam also traps silt and clay and stores water, altering the amount of water and oxygen gas present in the underlying sediment. This controls the types of bacteria and plants that can survive in and beneath the beaver pond. Microbes living in the bed sediment of the pond can transform and store much greater quantities of nitrogen—about a thousand times more—than can microbes living in the more oxygen-rich sediment beneath a riffle in the stream. Bacteria living in a wetland can remove as much as 80 percent of the nitrate entering the wetland. The pond behind the beaver dam also stores dead plant tissue that contains nitrogen but decays only very slowly within the pond. Through these mechanisms, the beaver meadow accumulates nitrogen and holds it tightly, releasing very little downstream or back to the atmosphere.

In contrast, the adjacent upland soils leak nitrogen to the soil water, groundwater, rivers, and the atmosphere. The upland soils are much thinner and support fewer microbes. The presence of oxygen gas in these drier soils also results in a different community of soil bacteria and a different set of chemical reactions, resulting in much less retention of nitrogen.

Nitrogen falling from the sky is a major pollutant of the seemingly pristine environments of Rocky Mountain National Park. Most of this nitrogen comes from the urban and agricultural areas at the eastern base of the mountains. If Rocky Mountain National Park could once again boast the numerous large beaver meadows historically present in the park, this would be an important step in reducing nitrogen pollution.

As goes nitrogen, so goes carbon. Like nitrogen, carbon is a fundamental building block of living organisms. Plants take up carbon dioxide gas (CO_2) during the process of photosynthesis and bind it into chemical compounds—a carbon skeleton—from which organic molecules such as carbohydrates can be built. Animals ingest the plants, and both plants and animals die or shed excess tissue—think of leaves falling from deciduous trees in autumn—releasing carbon to the soil. Soil microbes pass the carbon back and forth, in the process releasing it back to the atmosphere or dissolved in water as CO_2 or attaching the carbon to mineral grains that remain within the soil as a form of carbon storage. Comparatively little carbon is stored in relatively dry, oxygen-rich, upland soils. Instead, living plants are the major reservoir of carbon. Within the beaver meadow, however, large amounts of carbon can be stored in dead plant tissue and in the soil itself because decomposition and microbial activity resulting in release of carbon to the atmosphere or water are slowed in the saturated soil.

The North St. Vrain beaver meadow effectively stores carbon, nitrogen, and other nutrients as long as the meadow remains saturated as a result of the dams and other modifications by the beavers. In this sense the meadow halts hidden flows within the stream, continually taking in and releasing very little. In other respects, however, the beaver meadow extends the flow of the stream across the adjacent hillsides, subsidizing a broad region with habitat and food.

Cold, Wet Hot Spots of Biodiversity

Meadow is perhaps a misnomer, because the word conjures images of a grassy plain. Crossing the beaver meadow on North St. Vrain Creek, one finds a small, gravel-bed channel that dries late in the summer; then a sandy bench covered in river birch; an abandoned pond rimmed by steep-sided berms overgrown with willows; the mucky bed of the pond supporting sedges, cattails, and water lilies; a drier patch in which ground junipers and a small pine have taken root; a narrow beaver canal thick with algae through which aquatic insect larvae wriggle; an old dam, partly buried in sediment, from which sticks with beaver-gnawed ends protrude; a pond actively managed by the beaver, in which insects darting above the water surface are captured by frogs along the pond margins and by fish in the deeper water; the main channel of the stream, flowing swiftly over cobbles and small boulders, with twigs and pine needles collecting in the eddies and ouzels surfing the riffles; another sandy bench; an old pond, now nearly filled with sediment and forming a marshy wetland in which snipe build their nests and moose come to browse on plants; a thicket of willows following the slightly sinuous course of another abandoned dam, home to nesting warblers that eat the insects emerging from the stream and the ponds, and heavily browsed by elk and moose; a slow-flowing, secondary channel fed by upwelling hyporheic waters and lined by elephanthead plants whose flowers draw the bees; and finally a rising grassy slope favored by Rocky Mountain columbine that draw the rufous and broad-tailed hummingbirds. Square foot for square foot, there is no place in the national park with more diverse habitat than a beaver meadow.

You cannot swiftly cross this beaver meadow, but why would you want to? Crawling under densely intertwined willow branches or wading through a sticky-bottomed pond provides time to look around and take in the diverse plants and animals drawn to the biodiversity hot spot of the meadow. Seeds of birch and willow floating downstream lodge against the stream banks in the quiet water above a beaver dam or in the sheltered areas created by overhanging branches. Wetland plants can live in close proximity to upland species that like drier soil. The seeds of plants that prefer moisture germinate in different locations with time as the channels shift or a pond gradually fills with sediment and dries, creating successions of plant species. Beaver ponds host far more floating algae than flowing streams, and silt and clay trapped in the ponds and settling over the sand and cobbles flooring most of the meadow create a patchy pond-bed habitat that supports a more diverse community of bottom-dwelling invertebrates. The invertebrates and the variety of ponds in various stages of abandonment also support greater numbers and species diversity of fish than do segments of stream without a beaver meadow. A beaver pond sealed under a thick ceiling of ice is a particularly safe habitat for overwintering fish.

Although abundant water might seem a gift to plants in the dry climate of Rocky Mountain National Park, the superabundance of water in the beaver meadow presents special challenges. Hydrophytes, from the Greek *hydro* for water and *phyto* for plant, need distinctive adaptations to survive in the saturated soils of a beaver meadow. When water saturates soil, oxygen gas is depleted from the soil in a few hours to a few days as microbes and the roots of plants take up the oxygen. Dissolved oxygen is present in water, but this oxygen diffuses into the soil at rates a hundred times slower than oxygen is being consumed by the plants. Plant cells need molecules of ATP (adenosine triphosphate) to carry energy as the cells carry out their functions, but the generation of ATP is much more efficient when oxygen is present. Without some way to generate ATP in soil without oxygen, the amount of ATP in a root cell would support less than a minute of cellular activity, and the plant roots would die.

Hydrophyte plants cope with these challenges by changing the structure and location of their tissues and roots. First, they grow extra roots that can efficiently suck up whatever oxygen gas is available. Fast-growing secondary roots form off the primary root or off the plant stem near the waterline. These porous roots are more capable of absorbing oxygen gas and are located at least partly above the water level, maximizing the plant's ability to draw in oxygen. Second, the plants develop special aerenchyma (from the Greek words for air and infusion) tissue with large spaces between cells that allow oxygen to move more easily than in other plant tissue. Aerenchyma tissue also transports oxygen from the stem to the roots, helping provide oxygen to the beneficial microbes and fungi of the rhizosphere around the plant roots as oxygen diffuses out from the roots. Third, hydrophytes grow lenticels on their stems and roots. Lenticels are raised pores that facilitate gas exchange between the plant and its surroundings. Oxygen enters the xylem of the stem of hydrophyte trees through the leaves and, most importantly for the plant roots, through the lenticels.

Hydrophytes also undergo chemical changes that help them survive their watery environment. Waterlogging causes plant tissues below and immediately above the waterline to produce more of the chemical compound ethylene. Ethylene softens the cell walls and enhances aerenchyma tissue in the lenticels and the secondary roots. The presence of enlarged lenticels around the waterline allows increased oxygen movement into the stem. This in turn allows continued root growth and nutrient absorption even when the soil is saturated and depleted of oxygen. Essentially, by growing larger spaces within its tissues, a hydrophyte plant can accumulate and distribute oxygen more readily than other plants, keeping its submerged parts supplied with this vital ingredient.

Finally, hydrophyte plants adapt to the stresses of flowing current and fluctuating water levels by altering their morphology. They grow flexible stems and leaves so that, even when doubled over during a period of high snowmelt flow, the woody stems can once again rise tall and straight after the seasonal flood. And they spread their feet broadly. Willows, in particular, grow extensive, spreading root systems that allow them to remain structurally strong when the foundation in which they are rooted becomes soggy and yielding.

The distribution of individual plant species within the beaver meadow reflects the detailed history and characteristics of each spot of ground in which a seed germinates. Some hydrophytes are adapted to sites that are typically dry but periodically briefly covered by water. These plants must tolerate the boom-and-bust scenario of both flooding and drought. Other hydrophyte species prefer sites that are above the normal level of flooding but receive abundant water both through water movement within the soil and through periodic flooding. Plants in these regions of short- and long-term saturation include Rocky Mountain iris, buttercups, marsh marigold, and bistort, among herbaceous plants, and willows, birch, alder, cottonwoods, and spruce among woody plants.

A third group of plants can survive in the zone between the average annual high and low water levels. These plants include sedges, rushes, cattails, shrubby cinquefoil, and willows. The final group of hydrophytes lives in permanently flooded sites. Plants that are always flooded can be free-floating like duckweed, rooted to the bottom with leaves floating on the surface like pond lily, or rooted to the bottom with the vegetation fully submerged in the water.

Wetland plants near a beaver pond: elephanthead (*left*) and water sedge (*right*).

Clearly, an array of plants has evolved the necessary adaptations to thrive in saturated soils. Animals seldom found elsewhere in the national park seek these plants in beaver meadows. Wood frogs, which can tolerate freezing of their blood and other tissues, hibernate in the upper layers of soil and under leaf litter, emerging shortly after the snow melts. Moving during rain or evenings of high humidity in order to keep their skin moist, the frogs seek out ephemeral pools in which to breed. Their tadpoles eat algae, plant detritus, and the eggs and larvae of amphibians, including other wood frogs. Boreal chorus frogs and boreal toads also favor the beaver meadows, and increasingly rare boreal toads may depend on the meadows for survival. Tiger salamanders live in underground burrows in the forest, but return to water to breed, and they too may seek out the beaver meadows.

Consider a frog. Capable of long periods of absolute stillness, then abruptly bursting into an explosive leap many times its body length in order to escape danger. Changing from a tailed swimmer that resembles a fish to a four-legged jumper in the course of a few days. Frighteningly delicate and easily crushed, absolutely dependent on moisture and able to breathe through its skin. And yet, able to freeze each winter and revive the following spring. Vulnerability and resistance follow an interwoven dance through the animal's life.

Winter challenges every living organism in some way: food is limited, sunlight is reduced, and somehow living tissues must either be kept warm or shut down for the season. Poikilotherms—animals such as frogs, whose internal temperature varies significantly—typically shut down. The species of frogs in Rocky Mountain National Park have perfected the shutdown mechanism. Both species of frog present in the park, the wood frog and the boreal chorus frog, endure winter by freezing. These frogs can survive freezing of their internal body fluids to temperatures as low as −18°F in winter. Because of this, the frogs don't have to burrow deeply into insulating pond mud. Instead, they can hibernate beneath the surface of forest floor litter or downed logs where temperatures will certainly drop well below freezing.

The physiological mechanisms that allow a frog to survive such extremes read like the script of a science fiction movie. At sufficiently low temperatures, ice crystals begin to form within the frog. Hibernation sites must be moist, so frogs settled in for the winter are likely to be touched by ice crystals that start to form in the soil and leaf litter around the frog. Soon the crystals start to form under the frog's skin and in its abdominal cavity. A fast chill-down could kill the frog, but gradual freezing gives the body of the frog time to activate its protective responses.

Freezing of fluids surrounding cells disrupts cell function and causes death without protective measures. As the ice forms, chemicals dissolved in the body fluids are excluded from the growing ice and concentrated in the unfrozen fluids. This imbalance in the chemistry of fluids within and outside of the cells can draw water from the cells, causing cell shrinkage, destruction of cell membranes, and cell death. Frogs defend their cells with cryoprotectants.Cryoprotectants are like frog antifreeze: chemicals that eliminate the destructive effects of freezing. A freeze-tolerant frog stores large quantities of the cryoprotectant glycerol in its newly enlarged liver during autumn. Once ice starts to form in the frog's body, the glycerol is converted into glucose that accumulates within cells. Urea is also stored and functions as a cryoprotectant within cells. Glucose and urea limit cell dehydration and stabilize the cellular structure to limit cell damage. The release of latent heat as body water freezes allows the frog's cardiovascular system to continue to function for several hours, which provides time to distribute glucose and urea throughout the frog's body. Production of glycogen in the liver comes at the cost of other tissues in the frog, however. Body fat and muscles are chemically destroyed to create energy and simpler chemical compounds as part of the production of glycogen. In a sense, the frog's body destroys parts of itself to get ready for winter.

The body of a gradually freezing frog can also relocate substantial amounts of water from major organs to other spaces in the body. Some of the frog's organs can lose up to half of their water content during the first twenty-four hours of freezing. This limits mechanical damage caused by the growth of ice crystals during freezing of the organs. A freeze-tolerant frog may finish the freezing process with up to 70 percent of its body water in the form of ice, although much of this ice will be in places such as the frog's hind legs, rather than in the major organs. At this stage, the frog's limbs can no longer be extended. The animal does not breathe, and its blood flow and heartbeat cease. Suspended animation.

As temperatures warm and a frog's body starts to thaw, liquid water returns progressively to the frog's body fluids. Within a few hours, the heart resumes beating, blood starts to flow, and the frog breathes again. Imagine what spring means to a frog: resurrection.

Wood frogs in extremely cold Alaska exhibit these physiological changes more dramatically than wood frogs in moderately cold Ohio. This is the backstory of how frogs colonized high latitudes and altitudes as the Pleistocene glaciers retreated: they enhanced their cryoprotectant systems.

Now, however, these remarkable pioneers are declining in numbers within the national park and worldwide. A virulent fungus that seems to kill most species of frog has been spreading rapidly around the world since the start of the twenty-first century, and the continuing loss of wetland habitat further stresses frogs. Warming climate in Rocky Mountain National Park reduces the size and abundance of high-elevation wetlands, and the decline in beaver populations has largely eliminated beaver-created wetlands. Loss of wetlands means not only loss of habitat at the local scale but also loss of genetic connectivity among populations if individual frogs are not able to migrate overland to new wetlands and breed with others of their species. Now more than ever, the frogs and other water-loving plants and animals of Rocky Mountain National Park need the habitat supplied by beavers.

At the Heart of It All

As the frogs are thawing and resuming life, migratory birds return to the meadow for the breeding season. The beaver meadow provides each species what it needs. The abundance of insects feeds warblers. Waterfowl feed among the diverse aquatic plants in the beaver ponds. Red flowers attract hummingbird pollinators. Wading birds are drawn to the fish and amphibians in the ponds. Even the standing dead trees where a new beaver pond inundated and killed trees provide habitat for cavity-nesting birds. The abundant food for each of these types of birds draws them into the beaver meadow, and the diverse habitats in close proximity within the meadow fosters the presence of many different types of birds.

Animals that range widely come to the beaver meadow to graze or hunt. Elk, deer, moose, and rabbits seek out the rich and diverse forage in the meadow. Foxes, coyotes, and bobcats come to prey on the rodents and rabbits of the meadow. The only place I have seen river otters in the national park is the beaver meadow, where they prey on frogs and fish. Just as a coral reef forms a biodiversity hot spot within the ocean, so a beaver meadow forms a biodiversity hot spot along a river.

And what of the animals at the center of this abundance? A beaver is born fully furred and able to see, but the little animal gets lavish care from its parents and older siblings during the first weeks of life. Because the little ones are too buoyant to enter and leave the lodge via the underwater entrance, plant food is brought in for them and soiled plant bedding is taken out. When they have grown enough to venture out, their forays exploring the world around the lodge are supervised by an adult or yearling who remains nearby in case a tired kit needs to climb onto the older beaver to rest. Even young kits are agile swimmers, and they play with one another and the older beavers, gradually becoming familiar with their home pond and the best places to find food.

Winter is the most challenging time for beavers, as for many animals, and the family huddles in their lodge, each animal reducing its metabolism but still burning fat reserves. The beavers make periodic forays from the lodge to maintain openings in the pond ice, but their pleasure at ice-out in the spring is reflected in the exuberance of their swimming. Spring is also a time for work: the adult male, in particular, zealously marks the territory with small mounds of mud that he scent-marks to warn other beavers away.

Thanks to the need to avoid predators such as wolves, coyotes, and mountain lions, much of the beavers' activity is nocturnal. Even where such predation is now rare because people have eliminated the big predators, beavers seldom relax their vigilance, a behavior that ecologists describe as influenced by the ghosts of predators past. Beavers can sometimes be seen at midday, however, and more commonly in early morning and early evening.

People who make the effort to observe beavers carefully tend to be charmed by the animals. Individual beavers have distinctive personalities, and they are clearly affectionate and playful with each other. Their vocalizations can resemble those of human infants, and orphaned beavers bond with human caretakers. The famous tail slap is typically a sign of alarm or aggression in adults, but young beavers and some adults clearly use the tail slap as a form of play with other species. And, despite the famous productivity of a "busy beaver," the animals pace themselves, taking breaks and varying their tasks in a manner that human factory workers might envy. The adults and yearlings in a colony can transform a previously undammed reach of river with stunning efficiency, however, building multiple dams in a single season and transforming a relatively dry floodplain into a beaver meadow.

Besides creating unique niches and microhabitats for plants and animals, beaver meadows can provide refuges during times of stress. With their relatively abundant surface and subsurface water, beaver meadows in the mostly

A beaver harvesting willow stems along the Dall River in Alaska.

dry national park mediate the effects of drought and wildfires. When a for-
merly active beaver meadow is abandoned by the animals and becomes a dry
grassland, the valley bottom can burn, as occurred in Moraine Park during
the 2011 Fern Lake Fire. As climate in the region continues to grow warmer
and drier, active beaver meadows are likely to become increasingly important
refuges.

The growth and death of plants and the movements of animals create
hidden flows between the beaver meadow and the surroundings. Within the
meadow, riverside plants drop leaves and twigs into the streams and ponds,
adding terrestrial nutrients to the water. Aquatic insects that emerge from the
water as adults add freshwater nutrients to terrestrial ecosystems when the insects are eaten by spiders, birds, and bats. Ecologists comparing streams with
and without beaver activity find that beaver-enhanced streams have greater
numbers of insects, larger individuals of particular insect species, and a wider
variety of insects emerging, as well as more predators eating those insects.

Water flowing down the relatively straight line of North St. Vrain Creek
enters the beaver meadow and spreads into a maze of surface and subsurface pathways. Some of the water will not reemerge from the maze, instead

remaining underground or returning to the atmosphere. Most of the water will eventually emerge from the beaver meadow and resume its rapid journey toward the plains, but this water will have lost its load of sediment, organic matter, and dissolved nutrients, which remain stored in the meadow.

The beaver meadow is the center of a rich network of exchanges among atmosphere, surface, and subsurface; microbes, plants, and animals; stream channel and floodplain; upstream and downstream; and fresh water and uplands. Water, sediment, nutrients dissolved in water, and the flesh of plants and animals come and go throughout the network, maintaining a streamscape that is much greater than the sum of its parts. The beavers are ecosystem engineers, building and maintaining the physical structure in which the exchanges occur. In a sense, the beavers are also philanthropists, storing biological riches and sharing them with the greater community.

We need a national Beaver Day, just as we have Arbor Day to recognize the value of trees.

4
Old-Growth Subalpine Forest

As you drop down from the alpine tundra, it can be hard to see the forest for the trees. First come the isolated patches of krummholz, the dwarfed and windblown trees sculpted by the relentless winds near tree line. Continuing downward, the trees close ranks and grow tall, crowding into one another. Their tops still sway in the winter winds, but the traces of the wind appear in a forest structured by blowdowns that snap trunks or topple entire trees, rather than in the twisted shape of individual trees.

This is the forest of Engelmann spruce and lodgepole pine, as well as limber pine and subalpine fir. By the time the average age of the trees reaches two hundred years, the forest takes on the distinctive characteristics of old growth: large, old trees, but also many snags and fallen dead trees. The largest remaining stands of old growth lie in the subalpine zone, in the upper catchments of North St. Vrain Creek, Glacier Creek, and the Big Thompson River, far up the steep, rugged canyons.

Following the trail along Fern Creek, a tributary of the Big Thompson River, up to Odessa Lake provides an entry into old-growth subalpine forest. From a distance, the canopy appears impenetrable except where trees killed by mountain pine beetles stand barren of needles. At ground level, the interior of the subalpine forest is darker than the montane forest, with its widely spaced, reddish-brown ponderosa trunks. Where lodgepoles dominate the subalpine, their somberly colored grayish-brown trunks can grow closely if the stand is less than a century old. After the century mark is passed, some of the trees die, and the understory takes on a more open appearance, although still mostly shaded. The lodgepoles are straight-trunked, and their living branches cluster mostly in the upper third of the tree, so the interspersed spruce or fir tend to draw the eye. Both have needles that are shorter and darker green in color than the pines. The spruce needles are spiky, and, with branches that resist bending or breaking, the younger spruces are armored to ground level. Fir needles are a little softer, giving rise to the mnemonic "spruce is spiky, fir is friendly," but fir branches are also difficult to bend or break.

The understory comprises primarily ground-hugging shrubs such as common juniper, kinnikinnick, and Oregon grape. These, along with the relatively few species of trees, are easy to identify. The limber pines have upturned branches and flattened tops; the narrow-crowned spruce and fir have cones clustered in the upper branches; the lodgepole pines have longer needles and mostly dead branches until the upper portion of the tree. A few dark splotches or small, pale green tufts indicate lichens on the tree trunks and branches, but these are not particularly noticeable.

Despite their modest appearance, the lichens and other microorganisms within the forest canopy maintain the hidden flows of material from the atmosphere into the trees. Like their counterparts in the rhizosphere, the largely invisible canopy organisms consume nitrogen and other nutrients, transforming the elements into a chemical form that trees can use. As the trees age, die, and fall over, another community of microorganisms renders the nutrients contained in the tree available to other creatures in the forest. The hidden flows of the subalpine forest center on those from the wider world into the trees, through the trees, and thence back into the wider world. Microorganisms enable and regulate all these flows.

The Higher Spheres

The biodiversity in this seemingly simple forest is concealed. Splotches of variously colored lichens growing on the boulders protruding above the litter of needles on the forest floor hint at the diversity of lower plants. Some of them are not so low, however. The canopy here hosts communities hidden from human eyes. As with the rhizosphere of the tree roots, there are spheres of the upper levels of the forest that are vitally important to the forest ecosystem. Most encompassing is the phyllosphere (from Greek *phyllon* for leaf), which

A tuft of Usnea lichen on a tree branch in the subalpine forest.

includes all the aboveground portions of plants used by microorganisms for habitat. Within the phyllosphere are the caulosphere of stems (from the Greek *caulo* for stem), the dermosphere of bark (from Greek *derma* for skin), the phylloplane of leaves, the anthosphere of flowers (from Greek *anthos* for flower), and the carposphere of fruits (from Greek *karpos* for fruit). The microorganism communities of bacteria, fungi, and single-celled organisms divide the plant spheres among themselves. Some of them are beneficial to the host plant, others damage or kill the host. The majority have no detectable effect on the plant. You could think of the forest as the trees and the organisms that benefit or damage them, but you could equally well imagine the forest as just so much substrate for the abundance and diversity of microorganisms.

Slightly more visible within the canopy—but only slightly—are the cryptogamic epiphytes. Cryptogams (from Greek *kryptos* for hidden and *gameein* to marry) are plants that reproduce by spores, such as fungi and ferns. Epiphytes (from Greek *epi* for in addition and *phuton* for plant) are plants that grow on another plant but are not parasitic. Lichens, mosses, liverworts, and algae are among the cryptogamic epiphytes growing on trees in temperate forests. The subalpine forests of Rocky Mountain National Park do not host the luxuriantly flowering bromeliads of tropical rain forests or the lacy lichen

Lichens cover most of the surface of a boulder in the subalpine forest.

draperies of temperate rain forests, but the Rocky Mountain conifers do have their own epiphytes. Back in 1975, ecologist Larry Gough identified epiphytes in trees along the North St. Vrain Creek drainage at differing elevations. His study revealed forty-six species of lichens, two species of moss, and one species of fungus growing on the trunks of various trees.

Most of the epiphytes cluster on the north and west sides of the trees, where the prevailing winds bring a continual supply of moisture and nutrients. The individual species present on the trees change with elevation. In the subalpine forest, the greatest number of species live farther up on the trees, where they are not smothered in snow for months at a time. Among the trees, Douglas-fir and subalpine fir have particularly rich epiphyte colonies, probably because the firm bark of these trees provides a stable base for the smaller plants. For epiphytes, trying to grow on bark that flakes off the tree frequently is as bad as a person trying to build a house on the shifting sand of a barrier island. Smooth bark that sheds water is also less than ideal; a little roughness that can hold a raindrop for a while helps the epiphytes absorb the life-giving water.

The abundance and species diversity of lichens also vary with the structural complexity of the forest. If all the trees in a forest are the same age, they are likely to be about the same height and to have similarly shaped crowns, all of which create a uniform canopy. Old-growth forests that include both younger trees and snags typically have more abundant and diverse lichens because of the more diverse habitats available in the canopy—some epiphyte species require dead wood, for example, whereas others are less particular.

Lichens are themselves colonies in which algae or cyanobacteria live among filaments of multiple fungi species. The algae and cyanobacteria obtain energy through photosynthesis, which allows lichens to grow without drawing energy from the host plant. Moisture and nutrients are blowing in the wind.

Rain, dew, and humid air hydrate lichens and activate photosynthesis. Think of the lichens in the relatively dry subalpine forest of Rocky Mountain National Park as little patches of tissue perched on a branch or trunk high in the air, far from the moisture reservoir of the soil. The lichens need to absorb and store water whenever it is available. They can store water within their tissue or in tiny folds and depressions on their surface. Dewfall before dawn can be especially important in filling these tiny lichen water reservoirs.

Interactions between precipitation and lichens go both ways: moisture allows the lichens to photosynthesize, but the lichens also intercept precipitation falling on the canopy and alter the water chemistry. A forest canopy has a structure of increasing intricacy at progressively smaller scales, from the distribution of the branches of individual trees down to the details of needles, bark, and epiphytes. All aspects of this structure can influence the amount, distribution, and chemistry of precipitation that reaches the ground surface. Branches, leaves, and needles can intercept snow and rain falling on the forest, so that the precipitation evaporates back into the atmosphere before reaching the ground.

The precipitation that makes it to the ground through gaps and drips— the throughfall—has a different chemistry than the precipitation that fell on the canopy. Epiphytes are so many little chemical reactors. The hairlike organs used by epiphytic lichens to attach the body of the lichen colony to tree trunks, branches, and needles can trap particles from the atmosphere and eventually leach chemical compounds into the throughfall. Compounds such as nitrate settle on the epiphytes in a fine rain of dust even when the air is dry. Epiphytes also trap aerosols—suspended fine solid particles like clay or liquid droplets like fog. Any of these gifts from the air can be taken into the tissues of the epiphytes and leached in a different chemical form into the throughfall. Bits of bark, leaf, and stem caught on the epiphytes also decompose and leach chemicals into the throughfall.

Small details, such as the angle of branches and the roughness of tree bark, affect how long throughfall takes to move through the canopy. The slower the transit, the more opportunities that epiphytes have to alter the chemistry of the water.

The little epiphyte chemical reactors contribute to the greater effect of the forest canopy, where rain and snow are altered in ways that can benefit other life forms before the water reaches the ground surface. As I have noted in describing the hidden flows in every environment in the park, nitrogen is a limiting nutrient for many plants. This is worth emphasizing, because most plants simply cannot survive and grow without enough nitrogen. Until human use of nitrogen fertilizers on agricultural land and combustion of fossil fuels dramatically increased the amount of nitrogen available to plants in many environments, ecosystems evolved around the processes by which nitrogen could be extracted from the air, water, and soil in a form useful to plants.

Some nitrogen is supplied to plants in the subalpine forest by the fungi of the rhizosphere. Nitrogen within the lichen epiphytes also eventually becomes available to other plants when it leaches from the lichens or when bits of dead lichen fall to the forest floor and are consumed by microbes. Nitrogen is also obtained from the air via the microscopic colonies that live within a single pine needle. Limber pines in the Colorado Rockies host communities of bacteria within their needles, and those individual needles can have a long lifespan, remaining on the tree for more than ten years. This longevity provides sufficient time for an entire bacterial community to develop within each pine needle. These endophytic (within the plant) bacteria likely enter from the needle surface via the stomata, and once inside, they assist the tree by chemically altering atmospheric nitrogen into a form accessible to the tree.

The forest canopy is very efficient at capturing nitrogen blowing by on the wind, but this is not enough. Measurements in the subalpine spruce-fir canopy suggest that needles and branches, along with canopy lichens and microorganisms, manage to grab up to 90 percent of the nitrogen moving through the canopy in the air. However, this supplies only 10 to 15 percent of the nitrogen needed for tree growth. The remainder must come from the rhizosphere.

The epiphytic lichens of the canopy also influence the microfauna (the smallest of the small among animals) and macrofauna of the canopy. Lichens increase the small-scale complexity of habitat and the associated diversity of invertebrates such as terrestrial mites, spiders, and insects, which constitute the microfauna. These organisms are preyed on by songbirds—the macrofauna—which also use the epiphytic lichens as nest material. Like the trees, all these animals rely on the ability of the canopy epiphytes to extract enough food from the air to keep themselves and their tree hosts alive.

When a Tree Falls in the Forest
The communities of the canopy and their ability to snatch life from the passing winds are largely invisible from ground level. By autumn, the splashes of color provided by flowers and fungi are mostly gone; only the pale purple fleabanes continue to bloom in the forest along Fern Creek. Movements of wind and water create a distant background sound behind the soft *dee-dee* calls of chickadees and the rapid thuds of a woodpecker. Periodically, the seemingly outraged sounds of a chickaree erupt into the quiet. Sunshine slips down to spotlight individual branches and the perfect little Christmas tree shape of a fir sapling. Woodpeckers have flaked and pocked the bark of the lodgepole pines, and pine beetles have left spots of oozing sap, as though the trees had been peppered with buckshot. The powdery, pale tan, rocky soil is visible beneath the layer of fallen needles. Walking through this open understory should be easy, but the forest floor also hosts deadfall bristling with branches that remain remarkably hard to break or bend long after the tree has fallen. An old-growth montane forest has a lot of downed, dead wood. An old-growth subalpine forest has more.

Downed, dead wood, which forest ecologists describe with the uncomplimentary phrase coarse woody debris, contributes about half of the total biomass present above the ground in an old-growth subalpine forest in the national park. This wood contains up to 10 percent of the total nutrient pool of all the biomass in the forest. Although 10 percent does not sound like much, the concentrations of nitrogen, phosphorus, calcium, and sodium increase as the wood decays, providing an important, slowly available pool of nutrients that trees can use. My ecologist colleague Kate Dwire describes downed wood

Woodpeckers have excavated cavities and removed bark from this lodgepole pine.

as bacon strips on the forest floor, slowly feeding life all around them. Slow is the operative descriptor; fallen lodgepole pine wood can take 340 years to completely decay in the subalpine. Engelmann spruce can persist for 650 to 800 years. The higher the elevation of the forest, the greater the amount of coarse woody debris and the slower the decay rates.

Slower decay rates at higher elevations reflect the processes that progressively transform wood from the tissues of a living tree to the tissues of other organisms. Fragmentation starts the transformation. Woodpeckers hammer out chunks of wood and trees shatter when they fall. Even the chewing, ingesting, and excavating of invertebrates such as bark beetles help to increase the proportion of wood surface relative to the volume of wood—the invertebrates, in particular, create wood dust. Smaller fragments decay much faster than large pieces because there is more surface area for fungi to colonize. The invertebrates helping to fragment wood also attract foraging birds or even bears; a bear can dramatically accelerate the process of fragmentation when it tears apart a downed log. As the wood decays and becomes weaker, it settles and collapses inward. This settling increases contact with the soil and encourages more microbes, invertebrates, and vertebrates to move in to the decaying log.

By the time downed wood is physically collapsing, however, the microbes and invertebrates have been at work. Loss of wood mass can sometimes be a one-to-one conversion to fungal biomass. Fungi in the form of mushrooms, puffballs, molds, stinkhorns, rusts, smuts, soft rots, brown rots, white rots—increasingly unattractive names—colonize decaying wood. Joining the fungi are bacteria. Together, they either live on cell contents or degrade cell-wall components. Invertebrates—mostly insects—attack wood directly by eating the wood. The insects also introduce microbes that hasten wood decay. Microbes living within an insect's intestines can be expelled in the insect's feces and proceed to decompose wood directly on the new surfaces exposed by the insect's tunneling.

Fungal spores transported into the wood by insects join the attack on the wood. Wood-boring beetles, carpenter ants, and termites all love downed wood, and every species of insect that tunnels into bark or wood has its own array of predators, parasites, and scavengers. A beetle or a carpenter ant is like a miniature Trojan horse, introducing a host of organisms that hasten wood decay.

The presence of microbes and invertebrates provides the mechanism through which elevation exerts its influence. Fungi can survive below freezing, but they do best in the very pleasant temperature range of the mid-70s Fahrenheit. The activity of fungi and invertebrates declines at cold temperatures, so that very little wood decay occurs during winter in the subalpine. Extremely high and low moisture levels can also limit the activity of wood-decaying organisms, with low moisture more likely to limit summer activities in the subalpine forest. If only a few weeks each summer are suitable for organisms that promote wood decay, that explains why a fallen trunk can persist for eight hundred years on the forest floor.

In Praise of Dead Wood

It is just as well that fallen trees remain intact for hundreds of years: it would probably be easier to describe which organisms don't use coarse woody debris in a forest than to explain all the vital roles provided by dead trees. Once upon a time—a time not so very long ago—dead trees were viewed with distaste. Think of the connotations of the words once used to describe old-growth forests: senescent (deteriorating), inefficient, decadent, overmature. These words, used by foresters focused on timber production and harvest—as opposed to forest ecologists—reflected a common perception of old-growth forests as wasted opportunities to create useful timber in a young, rapidly growing stand of trees. Even the phrase "dead wood" has very negative connotations when used to refer to people in an organization, for example. (Why do we only call it dead wood in a forest or a human organization, and not when it is used in our architecture or furnishings?)

These perceptions of forests as timber-production factories date at least to Gifford Pinchot's nineteenth-century views of utilitarian forests, but Pinchot, the first head of the US Forest Service, was influenced by European foresters who had been harvesting their trees like crops for centuries. By the mid-twentieth century, attitudes of managing forests to maximize timber production had been taken in the United States to the extreme, viewing old-growth forest with a sort of moral outrage as an expression of arboreal rot that had to be eradicated and transformed into efficient tree crops. The relentless march of silvicultural progress finally stubbed its toe in the Pacific Northwest.

One outgrowth of the United Nation's establishment of an International Biological Programme between 1964 and 1974 was a call for the development of interdisciplinary teams of scientists who could study entire ecosystems, rather than isolated pieces. In connection with this challenge, the US National Science Foundation funded a group of scientists from Oregon State University, the University of Oregon, and the US Forest Service to study an old-growth forest in the H. J. Andrews Experimental Forest of western

Oregon. As recounted in Jon Luoma's engaging book, *The Hidden Forest*, this remarkable group of people set out to study old-growth forests before the forests disappeared. The research, which continues today, resulted in a series of amazing discoveries.

The Andrews scientists were the first to conduct research in the forest canopy—developing a canopy crane to facilitate access—and in the process they documented the critical role of canopy lichens in making nitrogen available to plants. Although European botanists had described fungi associated with plants roots during the nineteenth century, it was not until the Andrews research that scientists recognized the biochemical processes of the mycorrhizosphere and documented the importance of old growth in sustaining abundant and diverse species. They recognized the multiple beneficial influences of coarse woody debris and large wood in streams, and they fundamentally altered the way ecologists and society as a whole thought about old-growth forests. Ecologists now view old-growth forests as ecological treasure houses that must be protected and preserved. A brief online search reveals old-growth descriptions with positive connotations—virgin forest, forest primeval—as well as advocacy for the practice of proforestation, which aims to promote the eventual development of old-growth forests.

Today, some people may still regard old growth as unattractive, wasteful, and decaying, but that reflects their ignorance of a forest ecosystem, in which coarse woody debris is so much treasure available on the forest floor. We can start with plants. Downed dead wood first hosts the epiphytes that inhabited the living tree. As other lichens and mosses colonize the surface of the newly created coarse woody debris, they enhance the ability of the wood surface to retain seeds and dead needles shed from the canopy. The accumulation of this organic soil allows the seeds of vascular plants—ferns, flowers, seedling conifers—that grow on the forest floor to germinate on the downed wood.

Vascular plants send their roots into the bark and rotting wood to extract nutrients and water, or root in the mat of decaying smaller plant bits that accumulate on the downed wood. The concentration of nutrients and ability to hold water increase as the wood decays, but vascular plants require mycorrhizal fungi to really extract nutrients. Higher moisture levels in rotting wood can foster greater ectomycorrhizal activity than is present in surrounding soil. Pieces of downed wood can provide shaded microsites and protect plants from sediment moving downslope. Wood can also provide a leg (a

Cladonia lichens colonizing a decaying log.

root?) up for small seedlings that might otherwise be smothered by a deep snowpack on the forest floor. Many tree seedlings can grow on rotting wood, and some species germinate and survive better on dead wood. Gradually the wood fragments, as bark or pieces of inner wood slough off, individual plants topple from the log, falling trees or snags hit the log, or animals burrow into the wood or tear it apart in search of insects.

A wide variety of animals—amphibians, reptiles, birds, mammals—also use dead wood. Birds use snags for feeding, perching, and nesting sites. The list of cavity-nesters in North America is quite long. Multiple species of flicker, woodpecker, sapsucker, flycatcher, swallow, chickadee, titmouse, nuthatch, creeper, wren, and bluebird all need standing dead trees with cavities. Some species of bats also require snags.

Coarse woody debris characteristics such as piece size, decay state, wood species, and overall abundance of wood influence how animals use the downed dead wood. Less decayed wood provides perches and cover for runways. For a vole or a shrew trying to keep from becoming dinner for an owl, abundant downed wood can allow the animal to move through the otherwise exposed and hostile environment of the forest floor. Southern red-backed voles, for example, prefer areas in a forest in which coarse woody debris is present across the forest floor rather than widely scattered. The voles can take cover under the wood from boreal owls and pine martens and can also feed on the fruiting bodies of mycorrhizal fungi growing in the wood and accumulating nitrogen from the decaying tree tissue. Voles eat fungi and lichens but, like some people, they prefer truffles, some of which fruit mostly in rotten wood. The voles "repay" the fungi and truffles by dispersing fungal spores in their feces—the scurrying voles spread the spores of the minimally mobile fungus across a broader area of forest floor, opening up new colonization sites for the fungus. This is a highly interdependent and well-adjusted community: Voles need truffles for food. Truffles depend on voles for dispersal of spores and on a mycorrhizal tree host for energy. Trees require mycorrhizal fungi to obtain nutrients, and trees provide decaying wood needed by voles for cover.

Pine martens use downed wood in many ways beyond a food larder. The martens are active day and night year-round, but they do need to rest. During the warmth of summer, a marten commonly rests in the tree canopy. In winter, however, martens seek out old-growth forest so that they can use the downed wood for resting spots. The wood traps small air spaces and conducts heat or cold less readily than soil and rock. This allows the marten to warm the resting site with its body heat. Sites between the ground surface and the bottom of the snowpack and under or next to downed logs and stumps are the best. Mature adults are most likely to occupy these places, leaving the juveniles to get along as best they can in more marginal sites. The martens also rest for longer periods of time where wood forms part or all of the resting site, conserving energy for the all-important hunting of voles, mice, and shrews that also occupy the space at the base of the snowpack.

Small mammals excavate decayed wood, providing greater access to the wood for amphibians and reptiles. As wood decays, the invertebrates and the fruiting bodies of fungi promoting the decay provide food for vertebrates. There is a variety of food for vertebrates to choose from. Among the types of invertebrates found to use downed wood for food, shelter, and breeding sites are earthworms, slugs, snails, isopods, centipedes and millipedes, soil mites, bark beetles, wood-boring beetles, ants and termites, and bees and wasps. Then there are all the parasites and predators associated with these creatures.

Progressive wood decay creates opportunities for organisms to enter the log and use it for hiding spaces or thermal cover under loose bark. Invertebrates use downed wood as shelter in cold weather or as a hibernation site during winter. Beetles, for example, move from their summer home in leaf litter on the forest floor to winter homes under the bark of downed wood or tiny chambers excavated in the wood. The invertebrates, like the vertebrates, are in the wood for food and shelter. Invertebrates can eat nutrient-rich inner bark, less nutritious wood, or the fungi living on decaying wood. Carpenter ants, termites, and carpenter bees use coarse woody debris as nesting sites.

In total, scientists have documented vertebrates using coarse woody debris for cover, feeding, reproduction, resting, preening, and bedding, as well as sites for a lookout, communication (drumming), sunning, roosting, hibernating, food storage, and bridges. (While moving laboriously among downed, dead wood, I have often envied squirrels moving with swift ease on the bridges of tilted, fallen dead wood.) Any particular animal species may use dead wood for all, several, or just one of these purposes. Among small mammals—shrews, chipmunks, squirrels, woodrats, mice, voles, and ermine—the species using coarse woody debris average 70 to 99 percent of the total number of small mammal species in all kinds of forests.

No matter how you look at it, a forest is far poorer and supports much less life if old, dead, and decaying trees are absent. The presence of dead wood reroutes the hidden flows of nutrients in ways that support a far greater abundance and diversity of life than can exist without the dead wood. A single tree hosts an ecosystem, and the composition of that ecosystem changes as the tree changes from living to dying to dead to decaying. Once the dead tree falls, the snag community is replaced by the plants and animals hosted by coarse woody debris. With time, organisms using the surface of a hard dead log give way to organisms using the interior of a decayed log. Loose bark provides very small creatures a hiding space and shelter from extreme temperatures. Progressively softening wood is chewed, clawed, tunneled, and excavated by insects and small mammals, and successive invertebrates and fungi move into the new neighborhood. Hundreds of years after a tree has fallen, only a spongier feel to the soil, a line of saplings, or a linear shadow of slivers of decayed wood may indicate that here a fallen giant once lay, sustaining millions of other organisms during its long good night.

Beetle Bonanza

Despite the vital ecological roles of dead wood, too much of a good thing can be problematic. Walking through the subalpine forest highlights not only the amount of downed wood, but also the number of dying trees. The signs can be as subtle as tiny patches of beetle dust at the base of a spruce tree. The forest die-off associated with mountain pine beetles received a lot of attention in Colorado and across western North America at the very end of the twentieth century and the start of the new millennium. Widespread death of pine trees remains an obvious trend, but now the death of spruces is becoming increasingly apparent.

Another "tree killer" beetle, the spruce beetle, is the grim reaper for spruce. Like mountain pine beetles, spruce beetles are native bark beetles. Spruce beetles prefer Engelmann spruce but also sometimes bore into blue spruce. The beetles get around: they are present from Alaska and Newfoundland down to Arizona and New Mexico. In Colorado, they are mostly present in forests above 9,000 feet in elevation.

Under normal conditions, beetles bore mostly into downed trees. More downed trees after a wildfire, blowdown, or avalanche can increase habitat so substantially that beetle populations increase and start to bore into living trees. The beetles like their wood aged. They mostly attack mature and old-growth trees in slow-growing patches of forest dominated by spruce. When rising beetle populations create competition, however, even slender young spruces are selected by the insects.

Individual beetles live from one to three years. The adults, which are about the size of a grain of rice, emerge from dead or dying trees in late spring and early summer. Females then bore through the outer bark of another spruce and create intricate galleries in the sapwood. The problem for the tree is that these galleries disrupt the ability of the tree to transport nutrients within its tissues. Females lay eggs in their excavated galleries and the eggs hatch into tiny larvae that spend the winter growing in their woody homes. The growing larvae tunnel farther throughout the sapwood, sucking nutrients from the tree and eventually so disrupting its nutrient circulation system that the tree dies. About eighteen months after the initial invasion of the tree, the larvae turn into pupae. Some of the pupae overwinter in chambers at the end of the larval galleries. Others emerge from the host tree, only to bore back into winter quarters at the base of the tree, where woodpeckers are less likely to excavate and eat them and the accumulating snowpack insulates them from cold air temperatures. After two years, the beetles emerge from the host tree and look for a new host in which to start the lives of their descendants.

Why so many beetles now? As with mountain pine beetles, extremely cold winter temperatures can kill overwintering spruce beetles. But extremely cold winter temperatures are not what the Rocky Mountains have been experiencing during recent decades. On the contrary, warmer winters and drought-stressed trees are favoring beetle expansion at a rate that woodpeckers and other native beetle predators cannot sufficiently control. One of the defenses of a spruce tree is to exude sap that blocks the site where an insect is boring into the wood, but drought-stressed trees are less likely to be able to efficiently smother boring beetles in sap.

A study of spruce beetle migration patterns starting in 2017 revealed unexpectedly high numbers of the insects. The winter of 2018 was warmer and drier than the preceding winter. The beetles used the opportunity to emerge from the trees earlier in the season, continue to fly in search of new trees for a longer period, grow larger, and have a greater proportion of females in the population. The scientists monitoring the beetles extrapolated their results to infer that, as air temperatures continue to increase, beetles might reemerge and repeat their life cycle twice in a year, with predictable consequences for rates of tree death.

Records of fossil pollen and beetles reveal that extremely large beetle outbreaks have occurred periodically in the past, but with warming climate there really may be something new under the sun. Ecologists try to predict how ecosystems will change through time based on records of change in the past. Fossil pollen abundance in cores of lake sediment, for example, can reveal how the species abundance in forested areas has changed during past beetle outbreaks. Now, however, spruce trees killed by beetles may be replaced by a forest dominated by pine species that would have grown only at lower elevations a century ago.

All the other species in the forest are influenced by the age and composition of the trees. Spruce die-off is likely to ripple through the forest ecosystem in ways that enhance habitat for some species and harm others. Thus far, bees look like winners in this brave new forest. Death of the canopy trees allows more sunlight to reach the forest floor, increasing the abundance of flowering plants.

Jake Ivan and a group of Colorado wildlife biologists used wool soaked in peanut butter to attract animals to tree-mounted cameras during 2013 and 2014, and then used information about the twenty-six animal species record-

Aspen trunks scarred by elk, which eat the inner bark of the tree during winter.

ed in three hundred thousand photos to examine how tree die-off influenced animal presence. Elk, mule deer, and moose initially preferred sites where many trees had died, although the presence of moose declined with time. As with the bees happily feeding on more flowers, the greater sunlight reaching the forest floor supported new growth of saplings such as aspen, on which elk, deer, and moose browse. Chickarees were fewer in number than in adjacent forest with fewer beetle-killed trees, having lost their supply of conifer seeds. Ground squirrels and chipmunks were not as significantly affected because of their more diverse diets. Pine martens, black bears, and porcupines also seemed to be unaffected. Woodpeckers, as might be expected, did well as long as standing dead or dying trees remain present, but eventually their food sources declined as younger trees grew under the new openings in the canopy.

Ultimately, we do not know how changes in the subalpine spruce forest may affect many of the species that depend on this type of forest. Chickaree, pine marten, snowshoe hare, boreal owl, Clark's nutcracker, three-toed woodpecker, the boreal toads that live in wetlands among the high elevation spruce-fir forests and are one of Colorado's most at-risk amphibians, and the reintroduced Canada lynx—how will widespread die-off of subalpine spruce limit their ability to survive?

Consider the boreal owl. Small by owl standards, with a body only eight to eleven inches long, the boreal owl is dappled brown and white, with striking yellow eyes and a facial ruff that makes the bird look perennially startled. The owls are exquisitely adapted to nocturnal hunts for their favored prey of small rodents: one ear opening is high up on the skull and the other is much lower, helping them pinpoint the rustling on the forest floor that indicates a vole or mouse. The boreal owl's varied diet of small rodents and birds suggests that they could do well in a variety of environments, but the owls are cavity-nesters, and cavities are best excavated in dead trees. Dead trees will presumably always be present, and the owls should have an abundance of nesting sites as pines and spruce die in large numbers, but what happens when those trees fall and decades must pass before the trees that replace them are old enough to create snags suitable for cavity-nesters? In a balanced, functional forest ecosystem, boreal owls and other cavity-nesters could move to new nesting sites, but there may be a gap in time between the toppling of the trees currently dying so rapidly and the maturation and death of the trees that replace them. We do not know whether the owls will be able to survive such a gap, in part because the nineteenth-century history of deforestation in and around what is now the national park largely eliminated old-growth forest. Despite the apparently completely natural character of the national park, two centuries of sometimes intensive environmental alteration by people have narrowed

A chickaree (pine squirrel) on the forest floor near its midden.

the margins of survival for some species, including those that depend on old-growth forest.

Although subalpine forests on both the eastern and western sides of Rocky Mountain National Park are among the most affected regions of Colorado, the reach of the little spruce beetle is wide. Between 1996 and 2018, the beetles caused accelerated tree death in 1.84 million acres of forest in Colorado, with about 40 percent of the state's spruce-fir forests affected since 2000. The year of 2014 was the high-beetle mark thus far, with five hundred thousand acres attacked by beetles in that year alone. Across the Southern Rocky Mountains, the total since 1994 is four million acres. And then there is the spruce budworm, which killed approximately 131,000 acres of trees in 2018—this is not the best of times for old trees.

A forest, like any ecosystem, is more than the sum of its parts. The overarching flow within the subalpine forest is that of nutrients, snatched from air and soil by the nearly invisible workhorses of microbes, held for a time in the tissue of living trees, and then returned to air and soil and other organisms after the tree's death. When forests that require at least two hundred years to develop are abruptly dismantled by tree die-off via beetle populations that explode in a warming climate, organisms from the epiphytic lichens to the fungi of the rhizosphere are affected. Some version of the forest will continue into the future, but it may not resemble the forest we currently know.

5
Subalpine Lake

Loch Vale lies within a landscape of rock, on which trees and water make only faint impressions. From above, the lake looks tucked into a pocket of the valley, closely surrounded by steep rock walls on three sides. At lake level, the eastern side appears as a rim, with a void beyond. Tan intrusions stripe the pale gray granite walls or dip and backtrack in random swirls that record the violence of geologic history. Ice forming in small crevices has fissured the seemingly impenetrable rock, and water seeping down the rock face has left smears of black where microbes and lichens have colonized the moist surface. The cliff beside the eastern outlet has a rubble-strewn ramp, as though long ago someone started to blast a road into the rock wall and then abandoned the project. This ramp marks a zone of closely spaced fractures in the rock that have been blasted, but only by water freezing and melting, over and over, in millions of tiny cavities through a long succession of winters, each freezing cycle forcing the rock apart a little more. Lichens create a greenish-gray film over the fallen rocks.

Broken rock and water cascade down from a higher bowl that contains Lake of Glass. Icy Brook ties the lakes together, falling from Sky Pond to Lake of Glass and then the Loch in twisting ribbons of white that ice arrests in motion by early autumn. Long icicles tilt away from the vertical as a result of the steady wind while they were freezing. The icicles flicker as though lit from within when the trickle of water still flowing beneath them reflects the

A September view of Loch Vale in Rocky Mountain National Park, looking west to the Continental Divide.

weak sunlight. Bouquets of ice flowers bloom downslope from the icicles. Ice bulbed like sausage links stretches down the rock faces.

Many of the national park's subalpine streams follow a similar stepped path down through multiple lakes, with fingers of tributary streams stretched up to high points all around the lakes. Some of the streams emerge at the downslope end of snow or icefields. Others start as a small but steady up-welling of groundwater from the fractured bedrock. The streams become most noticeable during snowmelt, when the larger channels branch around logjams or small forested islands. By autumn, each creek shrinks back to its lowest flow, and the smaller channels are completely dry.

Wherever a depression has formed in the bedrock of the valley floor, wa-ter can pause and collect. In places this creates a squishy meadow such as the one downstream from Andrews Glacier near Loch Vale, or the shallowest possible pond, such as Embryo Lake upstream from the Loch. These depres-sions form where more densely spaced fractures in the bedrock cause the rock to weather more readily, enlarging the fractures and providing a conduit for groundwater to rise to the surface and join the stream. Deeper depressions at intervals along the valley reflect the track of a glacier during the ice ages. Al-

A tributary into Loch Vale, frozen by October.

pine glaciers flowing down valleys create regularly spaced depressions in the
rock, separated by steeper segments. When the ice melts, a lake forms in each
of the treads in this glacial staircase. Because of their resemblance to beads on
a rosary when these lakes are viewed on a map, they are known as paternoster
lakes. At some lakes, a moraine left as the glacier paused in its retreat forms a

low wall of sediment across the valley bottom and helps to pond water. Other lakes are formed in a bowl eroded into the bedrock. The highest paternoster lake in a sequence forms in the glacial cirque at the head of the valley and is known as a tarn, a term derived from an Old Norse word for lake.

In Rocky Mountain National Park, lakes cluster just below the Continental Divide on both the eastern and western sides. More than 150 lakes are present, mostly within the subalpine zone. The lowest natural lake in and around the park is Grand Lake at 8,300 feet elevation. The rapid decrease in precipitation with declining elevation limits the existence of lakes at lower elevations.

When I visit Loch Vale in October, a forceful wind rushing toward the east raises whitecaps on the lake. Waves thunk into little embayments in the lakeshore with a sound like someone dropping a large stone into deep water. Ice has begun to form along the edges of a larger embayment near the lake outlet. The ice is ridged into frozen ripples that are milky with bubbles. Long, delicate crystals bristle into the water beneath the ice. As the autumn advances, ice will seal the lake water beneath a translucent layer that appears waxy when the low-angle sunlight reflects from the ice surface. Long cracks frozen shut will scar the ice like lines on a checkerboard. Some of the broad plates outlined by these cracks will heave up or downward like a buckling sidewalk, while remaining frozen. Swirled textures on the ice surface make the ice seem alive, only temporarily arrested in motion by freezing.

The taller trees on the ridge crests are wind-combed, their tops bent and their branches twisted toward the east. Although this is an old-growth forest, the size of the trees does not rival those at lower elevations. The diameter of even the largest trees seldom exceeds two feet. Occasionally the wind blows a shoreline tree into the lake. Once submerged, the trunk can last for hundreds of years, creating cover for small fish and aquatic insects. As fungi and bacteria decompose the wood, the decaying log can provide a source of carbon for algae growing on it. Some of the wood entering the lake drifts to the outlet, forming a small floating jam. The jam shrinks during years of high outflow that carry some of the logs into the outlet stream, then grows during years when more wood enters the lake.

Transient sunlight plays with the scenery, briefly highlighting a tree here or a rock face there, never bringing the entire lake into the warmth of full color. The lake itself looks impenetrable, the depths hidden in darkness despite the clarity of the water. In reality, the lake is hardly unfathomable. Water depth averages about five feet, reaching a maximum of fifteen feet near the outlet.

Visually, a lake is easy to delineate. Water enters at the inlet and flows beyond the lake at the outlet. The lake is bounded by shores, a water surface, and a bed. But, like a stream channel, these visual boundaries are porous, with continual exchanges across boundaries between the lake and the greater environment. The materials crossing these boundaries—water, dissolved chemicals, sediment, organic matter, organisms—have undergone transformations before they enter the lake, and the lake further transforms them.

If our vision could detect microscopic movements and energy at wavelengths beyond the visible spectrum, the lake waters that we commonly describe as still would instead appear to be as busy as our most densely populated cities. Windblown particles of silt and clay sift gently down through the water column. Microscopic plants—phytoplankton—convert sunlight to plant tissue. Microscopic zooplankton eat the phytoplankton. Larval caddisflies safely ensconced in their form-fitting cases trundle about the lake bed in the shallows as they search for bits of organic matter to eat. Water seeping into the lake underground brings dissolved minerals from the surrounding forest soils and gradually melting frozen ground. At most, we're likely to see the caddisfly larvae or the occasional trout, but far more is happening within the lake than meets the eye.

Isotopic Fingerprints

All the water entering Loch Vale originates as precipitation, but the paths followed by precipitation make all the difference. Seventy percent of the water entering the lake starts as moisture evaporated from the Pacific Ocean, carried eastward across the coastal mountains and individual ranges of the Rocky Mountains before being dropped as snow. Some of the snow falls outside the lake's catchment but is then redistributed by the strong winter winds. Snow accumulates on the surrounding rocky or forested slopes and, in flowing down to the lake as meltwater, dissolves chemical compounds from the rock, ice, and soil. Older precipitation that has been held in soil or fractures in the rock is pushed out by newly arrived water and moves down into the lake carrying chemicals dissolved from the soil and rock. Ancient precipitation frozen into glaciers, icefields, and talus slopes melts and carries yet more chemical compounds into the lake.

Rain falls directly onto the lake. Most of the rain comes from the east, having passed over the Great Plains and the urban and agricultural lands immediately east of the mountains. Nitrate and other materials brought in by these rains reflect the routes followed by the moist air masses from which the rain is wrung.

The isotopic composition of the water tells the history of the water's movement. Isotopes are varieties of an element that have different numbers of neutrons, and therefore different mass, but the same chemical properties. Water is a simple molecule, composed only of oxygen and hydrogen, but both oxygen and hydrogen have isotopes. Hydrologists use the ratios of oxygen (oxygen-16 and oxygen-18) and hydrogen (hydrogen-1 and hydrogen-2) isotopes to infer two aspects of water movement. The first involves atmospheric transport and the second involves movement once the water falls as precipitation and moves through the ground.

Water vapor carried in the atmosphere comes primarily from the tropical regions of the oceans, where high water temperatures promote evaporation from the sea surface. The initial isotopic composition of this water changes as the water vapor moves within the atmosphere. Precipitation falling in the tropics is high in oxygen-18 and hydrogen-2. As the water vapor is transported farther from the low latitudes, these heavy isotopes are gradually rained out. Air masses moving poleward progressively cool and lose water vapor to precipitation. Moving over a continent or a mountain range and swirling into a convective storm hastens this isotopic separation. Matters are complicated by the continued addition of new water to the air mass as a result of evaporation or water release from the stomata of plants outside the tropics, so the detailed isotopic signature of precipitation at any place and time reflects the combined effects of different water sources and atmospheric travel paths.

Consistent patterns occur, however. The amounts of oxygen-18 and hydrogen-2 decrease from the coasts to the continental interior of the United States along dominant atmospheric flow lines such as the path of the jet stream from west to east across the continent. Air masses and water vapor reaching the western United States via different routes display different isotopic signatures. Water vapor coming from southerly sources such as the Gulf of California or the Gulf of Mexico has more of the heavy isotopes—hydrogen-2 and oxygen-18—than water vapor transported in from the North Pacific. Maps of the isotopic signature of precipitation across the continent clearly show the signature of more northern sources in winter and southerly sources during summer. The movement of air masses over mountains also leaves a clear signature in the water vapor, with gradual loss of oxygen-18 and hydrogen-2 at higher elevations.

Meteorologists can use radar and satellite data to track the movement of moisture-bearing air masses in real time. Like Sherlock Holmes reconstructing past events from bits of evidence, hydrologists and geochemists can track past movements of water using the isotopic signature of the water.

Further chemical change occurs in the water as precipitation infiltrates and moves through the subsurface, but now different elements are of interest. Because oxygen and hydrogen isotopic ratios do not change during surface or subsurface flow, these ratios cannot be used to distinguish new precipitation from water returning to a stream or lake after spending time underground. Water in the subsurface picks up chemical tracers such as dissolved sodium or silica—the details depend on the geology of a particular site and what the subsurface water moves through. Rainfall and snowmelt that infiltrate all the way to the groundwater eventually return to the surface with different tracer signatures than water that infiltrates only into the shallow soil. These differences allow hydrologists to use water chemistry to estimate the proportion of water coming from different sources. Seemingly simple, clear water carries the history of its journeys in the details of its chemistry.

More than half of the stream flow entering Loch Vale comes from snowmelt, a little less comes from subsurface water, and only a small proportion comes directly from rainfall. The proportion of water entering a stream or lake as subsurface water depends strongly on the steepness and material of the ground surface. Where the slopes are gentler and hold more loose rock and soil, more water moves through the subsurface. Where bedrock is exposed, much of the water runs quickly off the rock surface. Water from snowmelt is usually more acidic than subsurface water. As water flows through the soil or rock, chemical reactions buffer the acidity of the snow that is caused by atmospheric deposition of other materials such as nitrogen. Soils and boulder fields disproportionately influence water chemistry relative to their abundance. Although these features are relatively thin and have little storage capacity, fast geochemical processes occurring over hours to days can change water chemistry and buffer the acidic water released from melting snow. Snowmelt also displaces shallow groundwater that has been stored since the previous summer and that has therefore had more time to chemically react with the surrounding material.

The story of how the chemistry of rainfall and melting snow changes as water moves down to the lake is complicated enough, but the story also changes as the summer progresses. Acidic meltwater from the snowpack is most concentrated early in the runoff season. As snowmelt mixes with and flushes groundwater stored from the previous year, the concentration of acids declines. By late summer, most of the snow is gone and groundwater that has been stored underground for longer periods of time contributes more minerals that dissolve only slowly into water. During the summer, snowmelt occurs earliest at the lowest elevations and moves upward in elevation with time, so the patterns of water chemistry also reflect the timing of snowmelt at various elevations above the lake. Like sunlight moving up a slope under passing clouds, snowmelt and acidic meltwater move upslope through the summer, followed by less acidic groundwater seeping into the streams.

Loch Vale and Icy Brook record these diverse contributions in the volume of their waters. Maximum flow in June reflects snowmelt. Each day, the greatest solar radiation and warmest air temperature occur in the afternoon, but it takes several hours for the newly melted snow to travel down to and through the lake. Midnight comes before the flow at the lake outlet reflects the afternoon heat. By mid-July the midnight rush of outflow is gone, but large rainstorms can cause a quick, short increase in outflow. Less water flows from the lake in August, with the complete disappearance of lingering snow. During September, even large rainfalls do not create a spike in outflow from the lake; instead, these rains replenish soil moisture. Only the steady seep of water from relatively dry soils and fractured bedrock maintains a small outflow from Loch Vale during autumn.

The annual cycle of any subalpine lake pivots around snowmelt. When the snows come hurrying from the summits, they bring a lot more than just water. Dissolved in the water are the materials that have fallen onto the snowpack throughout the winter. Sulphate, nitrate, mercury, and other compounds attached to snow crystals or falling to the ground on their own move down through the catchment as the snow melts. The meltwaters flush organic carbon, aluminum, potassium, and compounds of phosphorus and nitrogen accumulated in the soil during the previous growing season, as well as elements dissolved from the underlying bedrock. A rich chemical broth enters

Loch Vale during spring and early summer, and this flush overwhelms processes within the lake that affect water chemistry. Only the bottom waters of the lake retain the chemical signature of winter. Essentially, the lake's chemistry is overwhelmed by the rush of water from terrestrial and subsurface environments.

As snowmelt ends, the lake again takes control of its own chemistry. Biochemical transformations within Loch Vale exert a progressively stronger influence on the chemistry of the lake water, becoming dominant when winter once more covers the lake with ice and limits incoming stream flow. The sources of nutrients used by lake organisms follow a seasonal seesaw: Dissolved organic carbon originates mainly from terrestrial plants and soils during snowmelt, while in winter most of the carbon comes from decaying phytoplankton in the lake. Nitrogen in the form of ammonium is barely present within lake water during the summer. Living organisms prefer to take their nitrogen in the form of ammonium or nitrate, so incoming ammonium is rapidly taken up by bacteria in the lake during the summer season of growth and biological abundance. Concentrations of nitrate, in contrast, remain high. The difference in abundances of ammonium and nitrate are the story of Loch Vale in the wider world.

The Critical Load

River catchments on the eastern side of Rocky Mountain National Park have a nitrate problem. Or, rather, the urban corridor stretching north and south of Denver creates a nitrate problem that shows up in east-side catchments of the park. As I've emphasized over and over, nitrogen is a vital nutrient for living organisms, but there is a limit to how much nitrogen any ecosystem can use.

Most nitrogen in a terrestrial ecosystem is contained in soil organic material and plant tissue. A small amount of nitrogen cycles annually between soil and plants, but some nitrogen is lost during this cycling, even in ecosystems that are strongly limited by the amount of nitrogen available. On the eastern side of Rocky Mountain National Park, this slow drip of lost nitrogen has become a gush. High-elevation streams in the park contain high levels of nitrogen throughout the year because the nitrogen leaks into the streams from the adjacent forests and soils. This leaking of normally precious nitrogen suggests either that forests, lakes, and streams are saturated with nitrogen or that it is moving through the ecosystem too fast for bacteria to process it. Or both.

How does nitrogen zip through an ecosystem? Much of the nitrogen entering the park's high elevation ecosystems comes from the atmosphere. Rain and snow carry nitrate. Ammonium and microscopic particles of nitrate fall from the sky even during dry periods. Upslope winds coming from the cities and farms east of the national park carry large amounts of nitrate released from fossil fuel combustion and agricultural fertilizers. This nitrate settles onto the catchment of Loch Vale, creating what might seem to be a gift for the alpine tundra and subalpine forest above and around the lake.

Human-generated nitrate emissions to the east of the national park have increased substantially since 1950, a trend that mirrors the rest of the world. Unfortunately, Loch Vale's catchment is poorly equipped to handle this gift from the sky. First, the timing is wrong. Much of the nitrogen is released from the snowpack early in the melt season before the growing season really gets under way and plants need nitrogen. Second, alpine and subalpine soils are thin. Much of the terrain above the lake is simply too steep to retain soil. Glacial ice covered the lake and flatter areas of the valley until about 14,000 years ago, and 14,000 years is not a very long time to form soils when each year includes only a brief period warm enough for the plant growth and microbial activity that form soils. Soils harbor bacteria that can make nitrogen available to plants, but meltwater moves rapidly through or across the thin soils, and the growing season during which bacteria are active is short at this high elevation.

Third, plants cannot live by nitrogen alone. Terrestrial plants also live and die by constraints imposed by temperature, precipitation, and the availability of other elements such as carbon. If any of these factors are limited, the plants will not be able to take up all available nitrogen. Fourth, alpine and subalpine plants have evolved to be thrifty with respect to nitrogen. Their metabolism is adapted to low levels of nitrogen, and they cannot shift gears and take up large quantities of this nutrient. Finally, much of the catchment above Loch Vale doesn't have any plants. The great majority of the land surface draining to the lake is bedrock and talus, and these surfaces do nothing to retain or transform nitrogen from the atmosphere.

The excess nitrogen spilling into the lake from the surrounding terrestrial ecosystem also overwhelms the biological processing capacity of the lake. Just

Loch Vale in winter. The wind has blown large bands of ice clear of snow.

as water must remain in soil for some minimum period of time in order for bacteria to assimilate the nitrogen, so water must remain within the lake long enough to facilitate nitrogen consumption by lake bacteria. Water resides in the lake for periods longer than a few days only during the winter, when microbial communities in the lake-bottom sediment can slightly reduce nitrogen content. Phytoplankton can also remain active beneath the ice and consume a little nitrogen thanks to fierce winter winds that blow the snow cover off the lake and allow the phytoplankton to photosynthesize. As in the terrestrial ecosystem, limited quantities of other nutrients such as phosphorus can put a brake on the abundance of algae and how much nitrogen can be assimilated.

When there is simply too much nitrogen, the species present in lake algal communities may change to favor organisms that can take advantage of the excess, or the lake may become acidic. Both changes are occurring on the eastern side of Rocky Mountain National Park. Lake nitrate concentrations there are significantly higher than in lakes on the western side of the Continental Divide and elsewhere in the western United States.

The sand and organic-rich black mud on the lake bed record changes in the Loch Vale ecosystem through time. Among the layers of sediment are the microscopic skeletons that single-celled algae known as diatoms create from silica dissolved in the lake water. The skeletons are works of art—tiny, delicately patterned spheres, diamonds, six-pointed stars, and spindles that settle to the lake floor once the diatom dies. Each species of diatom thrives under particular conditions of water temperature and chemistry, and each species creates a distinctive skeletal shape. The tiny skeletons among the sand and mud particles on the lake floor can be used to interpret the species composition and abundance of diatoms in the lake through time. Between 1950 and 1964, diatom communities in the lake abruptly changed from those typical of nitrogen-limited lakes to species associated with greater available nitrogen.

The average amount of nitrogen deposited in Loch Vale's catchment with precipitation during 1950 to 1964 was about three pounds of nitrogen for every two and a half acres of ground each year. Because this rate of deposition corresponds to a change in the composition of diatom communities within the lake, this rate defines the critical load—the amount of one or more pollutants that an ecosystem can safely absorb before its ecosystem function or ecological communities change.

The lake appears to have crossed a threshold of ecological change at about the year 2000, just as the global calendar passed into a new millennium. Since then, nitrate concentrations in the streams entering and leaving Loch Vale have been 50 percent higher than they were just a decade before 2000, even though atmospheric nitrogen deposition did not increase during that time. Did the catchment become saturated with nitrogen, or were other factors contributing?

One change influencing nitrogen levels in the lakes and streams is the air temperature. Summer and fall temperatures in the catchment of Loch Vale have steadily increased for decades. Warmer air accelerates the melting of ice in glaciers and rock glaciers in the upper portion of the lake's catchment. Grains of sand, silt, and clay with attached nitrogen are frozen into the glacial ice and released with ice melt. The meltwater coming from the glaciers also flushes nitrogen from soils hosting microbial communities beyond the glacier front: a double whammy that increases nitrogen fluxes to streams and the lake. And then there is the dryness.

Annual precipitation has been lower than average since 2000. By mid-summer, once the snow has melted and plants are steadily releasing water back to the atmosphere, soils dry out. Inadequate soil moisture limits soil microbial activity and the uptake of nitrogen by forests. Only the once-common summer afternoon rainstorms can replenish the soil moisture, but these storms are becoming less frequent.

Loch Vale and the other subalpine lakes on the eastern side of the park are experiencing not so much a double whammy as an avalanche of changes. Atmospheric deposition and melting ice increase nitrogen inputs, even as the warming, drying climate restricts the ability of forests and tundra around the lake to take up nitrogen during the growing season. Climate models indicate continued warming and drying for this region of the world.

Ice Bubbles and the Secret Life of Lakes

Phytoplankton, the microscopic community that powers the life of the lake, reflect these seasonal and longer-term changes in nitrogen. Named from the Greek words for wandering or drifting plant, the community of phytoplankton includes cyanobacteria, algae, and diatoms. These organisms all power themselves through photosynthesis and thus rely on the upper, sunlit layer of the lake water. They also require nitrogen and other nutrients. High-elevation lakes typically have fewer phytoplankton than lakes at lower elevations because of limited nutrients and cold temperatures. The ever-present wind brings algal spores to subalpine lakes, but only the toughest—those able to withstand low temperatures and the extreme seasonal fluctuations in sunlight and meltwater fluxes—survive.

The survivors can grow quickly, and phytoplankton have their seasons, just like the summer flowers and golden autumn aspen leaves of the forest surrounding Loch Vale. Phytoplankton populations increase when the annual snowmelt flush enters the lake. Diatoms and green algae dominate the community during this period, when water temperatures are cool and water chemistry reflects that of the incoming snowmelt. Phytoplankton populations drop to a minimum in midsummer, probably as a result of grazing by zooplankton, the microscopic animals that eat phytoplankton. Growth in the zooplankton community lags behind the phytoplankton population boom each year, but then gets to work consuming the tiny plants.

Water temperatures reach their maximum during August or September when snowmelt has finished and sunlight warms the water. The water of Loch Vale mixes vertically because of the surface agitation caused by wind. Blue-green algae come into their own in autumn, causing a secondary population peak in the phytoplankton community.

Phytoplankton population peaks again in early winter. As Loch Vale freezes, blasting winter winds keep the ice surface clear of snow, allowing sunlight to penetrate and supporting phytoplankton abundances comparable to those of the summer peak. The timing of ice formation varies from year to year, depending on air temperature and on wind, but the lake begins to freeze in October and is completely ice-covered by mid-November. The early, transparent ice comes solely from freezing lake water. With time, additions of snow cause the ice to become translucent, and by February it can be three feet thick. Although wind mostly keeps the ice clear of snow, the weight of drifting snow can buckle the ice in places.

Stream flow into and out of Loch Vale stops by the end of December, and the lake becomes isolated, self-contained but for small amounts of seepage into the underlying sediment and rock. The winter phytoplankton benefit from an increased concentration of compounds dissolved in the lake water. As water freezes, dissolved chemicals are excluded from the ice, increasing their concentration in the remaining liquid water. By midwinter, more than half the volume of lake water has frozen. The early winter population boom of phytoplankton goes bust relatively quickly, however, as sunlight disappears. The high rock walls surrounding the lake and the lower angle of the sun above the horizon combine to limit the light necessary for photosynthesis between November and February. Ice on Loch Vale can last up to 210 days, or more than half the year. With the return of longer hours of sunlight, the phytoplankton once again increase in abundance beneath the ice.

While the lake is freezing in autumn, bacteria continue their work of decomposition in the lake-bed sediments. The litter left behind by other living organisms—pieces of wood that float into the lake and then sink, wind-dropped leaves and pine needles, dead phytoplankton—is food for the bacteria. As living organisms, the bacteria also emit their own wastes in the form of methane bubbles. Methane bubbling up through the lake water diffuses into the atmosphere at the lake surface during summer. Come autumn, ice growing downward from the lake surface captures some of the

A Milky Way of bubbles and cracks within
the lake ice during spring.

bubbles, entombing them until
spring.

And then there is the wind.
Air blasting across the lake before ice forms transfers some
of the energy of its movement
to the water surface, creating
swirling vertical currents like
small water tornadoes in the
upper layer of the lake. These
currents align parallel to the
wind direction, forming alternating lines in which water and bubbles of
methane well up from lower layers, and "streaks" in which the water brought
to the surface moves sideways and then downward again.

Bacteria release their bubbles up toward the light. The wind stirs the bubbles, and the ice preserves them. By early spring, the ice covering the lake

A close view of a macroinvertebrate in the shallow water along a lake margin, here in Lac Bleu in the Swiss Alps. The larval insect is about two inches long, and the case it has constructed is elaborately decorated with plant fragments that stick out in all directions like feathers.

presents a fantastic mural. One portion contains flowers and mushrooms. Another is honeycombed with hexagonal cells of tiny bubbles. A swirl of arrested motion outlined in bubbles preserves a vertical current. Tiny tunnels stop just below the surface, as though the bubbles had burrowed upward toward freedom only to be stopped in the final fractions of an inch. A milky way of translucent bubbles speckles the clear darkness of the ice. Until it melts, the ice reveals the secret life of the lake.

The Lake Unseen

The spring increase in phytoplankton fuels the lake food web, starting with the zooplankton and continuing through the bottom-dwelling larval aquatic insects. Caddisflies and midges are particularly abundant in the shallows around the edges of Loch Vale. Like the species present in streams, the dead organic matter at the edges of the lake is consumed by collector-gatherers and collector-filterers that are in turn eaten by water beetles and by fish and, if the insects survive to emerge as winged adults, by fish and birds. Other macroinvertebrates such as mayflies, beetles, fingernail clams, blackflies, and stoneflies are less common in subalpine lakes in the national park, but they are present. From a human perspective, the predatory diving beetles that eat smaller things from tadpoles to larval midges have the most grotesque feeding strategy. The beetles bite their prey using their short, sharp mandibles, inject digestive enzymes into its body, and then suck the liquefied remains out of the hapless victim.

As they eat and churn through the lake-bed sediment in search of food, the aquatic insects control how bits of dead plants and animals move through the lake ecosystem or accumulate within the lake sediments. The macroinvertebrates also influence the abundance of fish within the once-fishless lake.

Scottish Lord Dunraven was entranced by the landscape of Estes Park when he first visited in the late nineteenth century. Despite earlier settlement and associated hunting of big game by farmers and ranchers, the hunting impressed Dunraven, and he set out to acquire as much land as he could. His intention to establish a private, European-style hunting preserve was thwarted when established settlers learned of his plans, but Dunraven stayed and pursued other dreams. Among these was the construction of a fish hatchery in Estes Park during the late 1890s. The hatchery comes with its own sagas, including the decimation of beaver populations in Lily Lake, as described by Enos Mills in his 1913 book *In Beaver World*.

One of the by-products of Dunraven's hatchery was the stocking of high-elevation lakes throughout what is now the national park. Greenback cutthroat trout native to lower elevations on the east side of the park were introduced to Loch Vale. From the 1920s to 1968, others followed in Dunraven's footsteps, introducing tens of thousands of Colorado River cutthroat trout native to the western side of the park, rainbow trout, and brook trout. Introductions have ceased, but Loch Vale now has a naturally reproducing population of trout that are mostly hybrids of greenback cutthroat and rainbow trout. These fish eat primarily larval and adult insects and spawn at the inlet and outlet of the lake.

The introduction of fish to Loch Vale started a cascade of changes that scientists have documented repeatedly when once-fishless lakes are stocked with fish. Fish alter the relationships between microscopic predator and prey by eating more of the larger species of zooplankton and some of the bottom-dwelling macroinvertebrates. Algae typically benefit from the reduction or

A subalpine meadow near the lake.

removal of the animals that eat algae. As phytoplankton populations increase, their millions of tiny bodies suspended in the lake diminish the clarity of the water. The diversity of plankton and macroinvertebrate species present in a lake declines when fish take up residence, and frogs also can decline or vanish entirely.

The phytoplankton that are now more abundant in Loch Vale thanks to the presence of fish appear only as murkier lake water. The fish themselves can sometimes be seen in shallow water along the lake margins, but most of the changes in living organisms and hidden flows within the lake remain unseen unless scientists measure differences in lakes with and without fish.

A Sensitive Bioreactor

Life in and around a subalpine lake is tenuous, but tenacious. Rock protrudes everywhere through the thin, pale soil and pokes up in bare domes among the wind-twisted trees. Many of the tree trunks are bent just above the ground, a record of the tree's effort to remain upright as the soil beneath it slowly moves downslope. Where water lingers in gently sloped portions of the valley a wet meadow can form, creating in summer a cushion of emerald green starred with yellow flowers.

Autumn leaves on the small aspens assume such a rich golden hue that they seem to glow with an inner light. Kinnikinnick on the forest floor turns yellow and the chickarees are too busy stashing seeds and cones to even vocalize. Chickadees and nuthatches can be heard during brief lulls in the ever-present wind. A gust of wind is made visible as a smear of silt and clay lifts from the beaten trail around the lake and rushes down the valley, the soil particles on their way once more. The wind topples a tree, but the noise of the moving wind overwhelms the resulting crash. Early ice at the edges of the lake is ridged by the wind, only the undersides of the ice displaying long, delicate crystals that finger downward into the water. Icy Brook is subdued, its secondary channels dry.

Loch Vale is a collector of tales told by the wind. Tales of tailpipes and feedlots emitting nitrogen far from the cold, high, windy basin of the lake. Tales of silt and clay from deserts to the west rushed over the mountain peaks by the relentless jet stream. Tales of floating algae grown from spores dropped by the wind; of fish brought by men and horses to a place they'd never been before; of insects whose parents pushed the boundaries of survival ever upward to new habitats.

Everything driving movement and change in this landscape comes from the atmosphere. Solar radiation streams down from outer space, providing the basis for photosynthesis. Streams of air bring moisture, nitrates and clay, and spores of algae. All of these incoming materials and energy go directly into the lake or react with the matrix of the landscape on the surrounding uplands—the rock, soil, streams, and vegetation—and then enter the lake. The materials mix and react in the lake, acted on by microbial communities and larger organisms, evaporating from the lake, seeping into the underlying bedrock, or flowing from the lake into Icy Brook.

The environmental media of rock, soil, and water are reaction chambers in which microbes transform and distribute the nutrients that support life and the toxins that restrict it. The reaction times and pathways differ among environments, as does the resilience of the environment to outside disturbances such as increases in nitrate deposition or the introduction of fish. The reaction chamber of Loch Vale absorbed decades of increased nitrogen deposition until the lake ecosystem was finally pushed beyond its critical load. Meticulous work by biogeochemists and ecologists has unraveled the deep and shallow histories of change within Loch Vale and pointed toward the lake's potential trajectory; these women and men have read the tales of the lake and guessed at its future. The intricate paths that connect past to future reveal both the sensitivity and the complexity of the lake as a bioreactor.

6
Alpine Flats

The tundra around Bighorn Flats feels like the roof of the world. The sky is endless and uninterrupted. Peaks rising 14,000 feet above sea level line the horizons, but the sky dwarfs them in a way that it does not when you look up to the peaks from lower elevations. Unlike the strenuous climbs up the steep creeks or hillsides elsewhere in the national park, here the walking is easy. Assuming you can breathe well in the thin air, portions of the Ute Trail or the Continental Divide Trail wind gently along undulating surfaces on which the absence of trees promotes long, long views. On a calm, sunny summer morning, the landscape might seem gentle, but this fleeting impression vanishes with the rising wind, rapidly coalescing clouds, and stunning claps of thunder.

In winter, there is no perception of gentleness. The wind blows endlessly, sculpting the snow into sharp-edged scallops and drifts, keeping large patches of ground bare throughout the season, even as adjacent drifts build to ten or more feet. Winter wind gusts can reach speeds of 200 miles per hour. Winter temperatures are COLD, and then colder, dropping as low as −21°F overnight, even though the intense sunlight of 12,000 feet elevation can melt snow and ice on south-facing surfaces at midday.

Hidden flows move slowly belowground, incrementally making soil from rock. The blasting winds create faster flows of nitrate and dust. Flows within the plants and animals of the tundra alternate between these extremes, de-

Wind-sculpted snow in the alpine zone of Rocky Mountain National Park.

clining to a trickle during the long months of winter and then accelerating to a rush during the brief summer.

Rock is not as obvious here as on the surrounding rocky peaks, but the tundra is predominantly a landscape of rock with a thin dusting of life on top. The fact that life is even present attests to the slow breakdown of rock that creates the skim of soil present in the tundra.

Cracking and the Critical Zone

How does something solid as a rock break down? It starts with subcritical cracking. Imagine tunneling down through the surface. Below the aboveground portion of the plants are their roots and the soil. Soil is mobile. Wind can lift soil particles and carry them onward. Snowmelt and rainfall can detach particles and carry them downslope, or dry particles can move downslope under the influence of gravity. Soil persists because under the soil layer is the weathering damage zone. This is where weathering—the breakdown of rock—damages the original mineral composition and rock structure.

Weathering has many tools. Ice crystals, plant roots, or secondary minerals that precipitate from solution in the soil expand into tiny cracks in the

Fractured rock in the alpine zone, Rocky Mountain National Park.

bedrock, forcing it apart. Water soaks into porous rock or moves through cracks, chemically reacting with the rock in an exchange of elements—a little hydrogen for the rock, a little calcium for the water, and soon the rock's mineral composition is no longer what it once was. The removal of overlying material via a landslide reduces the confining pressure and the rock expands and cracks.

Each crack is an entry point—a zone of weakness—at which foreign substances such as water can enter and alter the rock. As cracks propagate, the surface area exposed to chemical reactions increases and the movement of detached portions exposes new, fresher rock that undergoes chemical changes. Mechanical and chemical weathering work as a team to alter the bedrock.

Each type of rock has a breaking point. Once the critical stress exerted on the rock exceeds that strength, cracks will rapidly propagate throughout the rock. This is what happens when someone shatters a rock with a heavy blow from a sledgehammer. Such violence is not necessary to break rocks, however; tiny cracks grow slowly and steadily through rocks at much lower stresses both at the surface and underground. This is subcritical cracking. In either form of cracking, bonds within the rock are permanently broken by stresses that pull them apart with tension or tear them by shearing.

Microscopic cracks are present in all rocks within and between grains. At the scale of atoms, the mechanical propagation of cracks through the rock is helped by chemical reactions at the crack tip because cracks grow by breaking chemical bonds. This is a slow process, but there is all the time in the world for it to occur. Rock commonly spends hundreds of thousands of years within a few tens of feet of the surface, where subcritical cracking occurs.

The stress that cracks rock ultimately comes from three sources. The first is gravity acting on the weight of overlying rock or soil and the movements of that overlying material. These movements cause changes in the pressure bearing down on underlying rock. Second, tectonic stresses resulting from movements of plates at Earth's surface can crack rock. Third, closer to the surface, stress can result from freezing, precipitation of salt, growth of plant roots, and daily and seasonal heating and cooling of rock and soil, or the more dramatic heating associated with wildfire. All these stresses combine to cause subcritical cracking in rock from the surface down to about three hundred feet in depth. Starting a few feet down, the deeper portions of the zone of cracking and chemical alteration create a transition between intact bedrock and the upper, mobile soil. First the strength of the rock is diminished, then the rock is detached into the mobile layer, and finally the breakdown products of the rock move downslope under gravity or into the atmosphere on the wind.

All these processes occur within the layer of Earth's surface known as the critical zone. This is the permeable layer of the planet's surface, where bedrock is affected by water, air, and life, and where ecosystems are supported. The critical zone is fundamentally a layer in which rock is progressively damaged en route to the surface and eventually released into the sediment conveyor system that operates on hillslopes and in rivers. Earth's surface is littered with broken-down remnants of the planet's interior.

The critical zone is a dynamic interface between the magma and solid rock of Earth's interior and the atmospheric envelope of moisture and living organisms at the planet's surface. Upward surges of magma and energy from the interior build topography by raising mountains. Surges of atmospheric energy in the form of moisture driven by solar radiation tear apart the topography through erosion by glaciers, rivers, and rainfall. At any time and place, the balance between these forces creates a landscape that can be geologically transient—poised on the cusp of change if the balance between planetary and atmospheric energy shifts. In slow, steady changes and in abrupt, violent surges, the critical zone functions as a weathering engine in which rock is carried upward to be altered mechanically and chemically into forms that sustain living organisms.

Fire and Ice

The breakdown of rock is fundamentally influenced by the amount of moisture present: greater moisture equates to faster subcritical cracking and associated breakdown. This helps explain why rock breaks down so very slowly in the dry climate of the tundra in Rocky Mountain National Park. Just over three feet of precipitation falls each year, mostly in the form of snow, but much of this moisture goes right back into the atmosphere before becoming available to plants. Geologists estimate that rock is weathered and eroded from the alpine flats at rates of about sixteen vertical feet per million years. In contrast, the rate of rock weathering and erosion in the adjacent, lower elevation river valleys is nearly five hundred feet per million years. Rock in some portions of the alpine flats breaks down a little more readily, however.

Bedrock is heterogeneous. The size and chemical composition of mineral crystals within the rock, the alignment of crystals as a result of deformation within Earth's interior, and changes in grain size within sedimentary rocks all create irregularities that can be exacerbated by subcritical cracking. Consequently, portions of the rock tens to thousands of square feet in size can have more widely or closely spaced cracks. Where the cracks are more widely spaced, the rock better resists mechanical breakdown and chemical alterations. These tougher portions of rock can form high points known as tors in the alpine flats. The bare bedrock knobs of tors are commonly surrounded by blocks of rock that are up to three feet in diameter. The blocks have been peeled from the tors by ice crystallizing in cracks or they have been blasted off by lightning.

Global lightning maps suggest that sixty to ninety strikes hit each square mile of tundra in the national park every year. The electrical current created by a cloud-to-ground lightning strike can be anywhere from 10 kiloamps to 300 kiloamps (about a thousand times the energy used to light a football stadium). The result can be instantaneous heating of the ground surface up to 30,000°C, which is more than five times the surface temperature of the sun,

A boulder surrounded by shattered fragments, Rocky Mountain National Park.

during less than a millisecond. The instantaneous heating and expansion of air and moisture on and within the ground surface can blast rock apart at cracks and create unique tubes and lumps of rock by melting and fusing pre-existing minerals within the rock. Geologist Bob Anderson of the University of Colorado has a dramatic photograph of the results from a lightning strike that he witnessed on alpine flats. Dead sheep lying on their backs, their stiff legs pointed at the sky, litter the ground where a lightning strike instantly killed them and tossed their bodies six feet into the air. The lightning blasted a rent 150 feet long, four to eight inches wide, and two and four inches deep into the ground. That flow of energy wasn't particularly hidden.

Less dramatic but more continual are the cycles of water freezing and expanding in soil and cracks within rock. These cycles also shatter the rock of the tors and progressively lift larger stones within the soil to the surface in patterns of circles, nets, and stone lines. Cobbles and boulders are denser than the surrounding soil, so they conduct and retain heat differently. As soil freezes in the autumn, surface tension between water molecules can cause the water to migrate to spots where ice is already starting to form. The migration causes pockets of ice to form in the soil, and these pockets commonly underlie cobbles and boulders, like a little ice nest on which each rock rests. As water migrates to the ice nest during the period of soil freezing, the expansion of new ice crystals slightly lifts the overlying rock. When the soil thaws in spring, the ice nests melt more slowly than ice scattered throughout the surrounding soil. Moist soil oozes into the cavity beneath the rock once the ice nest does melt, preventing the rock from sinking back to its original level. This is how rocks get lifted toward the surface faster than the surrounding soil; a small increment each year, but enough over time to result in a concentration of rocks at the surface.

Rocks pushed to the surface by ice can create distinct patterns on the ground. In steeply sloping terrain, the rocks align into stone stripes down the hill. In gentler terrain, the rocks can be segregated into irregular circles that occur in groups resembling a large net when seen from above. Individual circles can have a diameter of a foot or two, or of hundreds of feet. The size partly depends on the intensity of freeze-thaw cycles and the amount of moisture in the soil that can be frozen into ice. Such patterns form only where the deeper portions of the soil remain frozen throughout the year. This lower layer of frozen soil prevents meltwater from percolating downward during spring and summer, so that the soil can be saturated even where little precipitation falls. On steeper slopes, the entire melted layer of soil can move slowly downslope each summer, forming subtle lobes and terraces that are a few inches to perhaps a foot tall.

As climate grew warmer ten thousand years ago, the deeper, perennially frozen soil in Rocky Mountain National Park thawed, and the stone lines and circles of the tundra stopped actively forming. A few patches of frozen ground lingered into the twentieth century, but rapidly warming air temperatures eventually destroyed them. Now that individual rocks are covered in lichens and partly sunken back into the ground, the patterns formed by the rocks can be hard to detect.

Ice dynamics at the smallest scales—freezing and thawing of water droplets and small pockets of ice in the soil—dominate the physical breakdown of rock. The ice facilitates subcritical cracking as well as the movement of broken rock by forcing the rock upward and then, as meltwater, lubricating the downslope creep of soil. The only real competitors in terms of bringing substantial amounts of gravel and smaller-sized soil particles to the surface are pocket gophers.

Northern pocket gophers are only six to ten inches long (a quarter of which is tail), but they move a lot of soil. As the only totally burrowing ani-

Example of active patterned ground in northern Sweden.

mal in the alpine zone, pocket gophers spend their lives underground, using their teeth and claws to dig an extensive network of tunnels from just below the surface to more than a foot deep. A single gopher can dig a tunnel up to a hundred feet long overnight. They remain active during winter, altering their summer diet of plant leaves and stems to primarily roots and tubers. Where a blanket of snow insulates the soil and keeps it from freezing, the gophers also continue to burrow, distributing plant seeds and aerating and then fertilizing the soil with their feces and urine. The soil they move can accumulate in the tunnels that the gophers maintain at the base of the snowpack. When the snow melts, these sinuous, linear mounds of soil sit on undisturbed ground until winds and rain smear them into more diffuse mounds. Careful measurements suggest that gophers can move enough soil to the surface, where wind and water erode it, to lower the ground surface by just over an inch per thousand years. Freeze-thaw cycles lower the tundra surface by about three and a half inches in the same time span, and wind and water erosion make a negligible contribution of hundredths of an inch. All of them—ice, gophers, wind and water—contribute to very slowly lowering the alpine flats.

For millions of years, processes ultimately driven by solar radiation have dominated the weathering engine of the alpine flats. Since the last period of mountain building ended about five million years ago, sun and ice have relentlessly cracked and broken the rock and fueled the growth of plants and animals that further disturb the cracked rock. Atmospheric processes also dominate rock breakdown and erosion in the adjacent lower terrains, but things have moved faster where alpine glaciers and rivers have sculpted the valleys, scooping the valley bottoms downward in a manner that has destabilized adjacent hillslopes and hastened landslides and debris flows. The alpine flats end sharply where a valley headwall drops steeply down to a glacial cirque. That steep drop represents a line where rates of weathering and erosion change abruptly from the slowly eroding flats to the rapidly eroding valleys. On the flats, the rapid changes come from the atmosphere.

Summer Frenzy

Wind rules the tundra in Rocky Mountain National Park. It brings the moisture that freezes to ice and cracks the rock. It brings the moisture that sustains plants, but also whisks that moisture away from the plants in desiccating blasts. It drops the temperature and limits the ability of plants to photosynthesize, of flying insects to move and pollinate flowers, and of warm-blooded animals to sustain their body temperatures. Wind brings the particles of silt and clay that are vital to forming soils capable of sustaining plants. In the alpine zone, the air is always in motion.

Only the hardiest and best adapted species can survive in the tundra, and they are few. Trees are not among them. Woody plants are present, but they do not grow tall. Photosynthesis cannot occur unless plant tissues are sufficiently warm. The short growing season of the tundra restricts the formation of new plant tissue, which is mostly directed into leaves and underground roots and rhizomes that store carbohydrates produced by the plant during the summer. There is not enough time and energy left to grow woody tissue and lift the plant far above the ground, nor is there enough water. So there is a tree line, above which trees cannot survive, and farther upslope there is a shrub line, beyond which not even ground-hugging woody shrubs can make it.

Average temperature in the month of July is a good indicator of just how tough the alpine environment is. As July temperature declines with increasing elevation, the number of plant species present on the tundra declines. The characteristics of the plants change, too. As the climate becomes harsher, the plants become shorter and more widely dispersed. Crowded plants can protect each other from wind, but patches of bare rock or soil separate plants growing under the most difficult conditions. Most of any alpine plant—up to 95 percent by mass—shelters underground in roots and rhizomes (from Greek *rhiza* for root and *rhizousthai* for take root). Rhizomes are horizontally growing underground stems that help the plant to spread with limited exposure to the cold air and wind. As a rhizome grows, new roots grow downward and new tufts of leaves grow upward, allowing what appears to be a separate plant to emerge from the ground some distance from the parent plant.

Besides a short season of growth, alpine plants must survive the intense cold of winter when the plant is dormant, as well as the occasional freezing period during summer. Remaining close to the ground not only limits wind exposure, but also allows the plant to take advantage of solar heat absorbed by the dark ground as soon as snow melts in the spring. Anything that gives the plant a jump on growth in the spring can make the difference between survival and death. Wintergreen or semievergreen leaves that develop late in the summer and do not wither during winter allow some species to start photosynthesis as soon as the weather is warm enough in spring (no delay to grow new leaves). A large proportion of alpine species contain anthocyanin pigments in their leaves and stems, creating a reddish color that is most obvious in spring before photosynthesis has kicked in fully. The anthocyanins result from carbohydrates stored in plant roots from the previous growing season. When rapid growth begins in spring, part of this carbohydrate reserve is diverted into new plant tissues and part of the reserve goes into anthocyanins that are capable of converting sunlight into heat to warm the plant tissues and enhance photosynthesis. The more anthocyanins present, the better the plant can withstand cold.

Rather than sheltering most of their tissue underground, lichens exist just above the surface. The tough outer layers of the fungus in a lichen protect the inner layers of algal cells enmeshed in fungal threads. Lichens have minimal requirements for survival and can begin photosynthesis as soon as the temperature exceeds 32°F, making them ideally suited for survival in the alpine zone.

Resist cold and bloom early—these are survival strategies for flowering alpine plants. Many alpine plants can start growing and start absorbing nutrients through their roots as soon as the temperature passes 32°F. This tem-

Plant at timberline on shores of Lawn Lake, Rocky Mountain National Park.

perature threshold is much lower than that of most lowland plants, which need warmer temperatures to "wake up" in the spring. Storing carbohydrates in roots and rhizomes also enables early plant growth in spring. While the plant is dormant during winter, these stored reserves remain untouched. Rapid growth in the spring depletes the reserves, but the plant restocks later in the summer. Flowering does not require air temperatures as warm as those needed to ripen seeds, so blooming early in the summer allows alpine plants to take advantage of summer's warmth for ripening seeds. The extreme version of this strategy, as practiced by mountain avens, is to grow flower buds late in the summer and retain the buds during winter so that the buds can open quickly when temperatures warm again.

Alpine plants must also survive wind and drought. Just as fur and feathers help insulate animal bodies against cold and wind, so hairs help protect plants from wind and dryness. Whether you are scanning across the tundra or examining an individual plant closely, the dominant color of the leaves and stems appears to be a pale olive green. In reality, the tundra species are as green as lowland plants, but their color is obscured by coatings of hairs. These hairs shield the plant's stomata and reduce water loss when the stomata are open. Hairy plants also create their own tiny greenhouse effect as the hairs trap heat and help warm the surface of the plant. The hairs can even diffuse the strong solar radiation present at high elevations, which can cause cell damage if combined with high reflection of light from lingering snow or ice. The paradoxes of sunlight are worth considering for a moment: just as humans need the vitamin D of sunlight but can damage their skin and eyes with too much exposure, so plants absolutely rely on sunlight for photosynthesis but intense exposure can damage their cells. Lacking our skin lotions, plants rely on hairs to deflect some of the light.

Another plant strategy for tundra survival is to stay slim. Grasses and sedges form a large proportion of the flowering plants present in the tundra. The narrow leaves and stems and the tiny flowers of grasses and sedges better resist the wind.

Perhaps the ultimate sheltering strategy, however, is to hide within the rocks at the surface of the tundra. Entire microbial ecosystems live in the tiny pore spaces within rocks. In the outermost layer of rock, these endolithic (from the Greek *endon* within and *lithos* stone) communities receive enough sunlight to rely on photosynthesis. Scientists began to pay attention to these communities in hot deserts during the 1960s and subsequently found them in other challenging places for life, including polar deserts and alpine ecosystems. Life in a rock can be good: protected from intense solar radiation, wild swings in temperature, and drying winds, yet able to obtain sunlight, water, and nutrients. Fungus, lichens, algae, and other microorganisms can be sufficiently concentrated to form a visible band of green just below the rock surface in the tundra.

Moss campion among rock fragments, in arctic Alaska.

Almost all the readily visible alpine plants in Rocky Mountain National Park are perennials with long-lived roots and rhizomes. Quick growth of long roots can help stabilize what little soil is present around the plant against erosion by wind and displacement during freezing and melting of ice. Moss campion may be the champion in this respect. A plant that is only a few inches across can send a taproot tunneling down into cracked bedrock to a depth of four or five feet.

Moss campion is a cushion plant. The ground-hugging, streamlined mound of a cushion plant minimizes wind exposure and maximizes the leaf surface available for photosynthesis. Thanks in part to the accumulation of dead leaves that absorb the sun's warmth, air temperature among the densely spaced stems of a cushion plant can be several degrees higher than the surroundings. The colder the outside air, the greater the difference in temperature. The short, dense branches of moss campion also trap and retain windblown soil particles. The alpine cold that slows plant growth also slows decomposition of dead plant tissue. Dead leaves remain attached to the stems of cushion plants, protecting the living leaves from abrasion by blowing snow crystals in winter and from drying wind in summer. The cushion plant thus recycles its own nutrients and catches what it can from the wind. These strategies make moss campion one of the more rapidly growing species on the tundra, although rapid is very relative here. The diameter of the cushion can reach half an inch in five years and seven inches in twenty-five years. The plant may not flower until it is ten years old and may not have abundant flowers until the mature age of twenty years.

Reproduction is a challenge for alpine plants because of the limited soil fertility and short season of growth. The blossoms of many alpine plants appear outsized among the diminutive stems and leaves. The flowers are actually not especially large, they just seem so in comparison to the rest of the plant. Once the flowers open in summer, they must attract pollinators. Although bumblebees are present in the tundra, flies are the most important pollinators. Insects with two wings—midges, crane flies, mosquitoes, hover flies, and gnats—all carry on the work of pollination in the tundra. Bees are immobilized below 50°F. Flies continue to move at lower temperatures, although they can be temporarily stilled on a summer day when the sun goes behind a cloud. But the flies have flowers for refuge, ideally white or yellow.

Air temperatures in some of the white or yellow bowl-shaped blossoms of the tundra can be several degrees warmer than the outside air as long as the sun shines, because the petals act like parabolic reflectors that focus the sun's rays and warmth onto the center of the flower. This not only hastens the development of pollen and fertilized seeds but also attracts insects that can rest in the warmth of the flower and distribute pollen among plants. For this little sunroom to function, the flower must keep turning to face the sun as the sun moves across the sky.

Flower stalks grow continuously during summer. Plant growth hormones produced in cells at the tip of a flower stalk seep downward to stimulate elongation in cells lower on the stalk. Sunlight inhibits the formation of the hormones, so growth is slower on the sunlit side of the stalk and cells on the shaded side elongate more rapidly. This intricate manipulation of hormones allows the flower to follow the sun each day.

Fertilized, mature seeds must also be dispersed. The omnipresent wind helps in this. Round, smooth seeds often end their travels in a snowdrift, along with windblown dust that can help provide a tiny pocket of soil in which the seeds can germinate when the snow melts. Plumed seeds can be dispersed by more gentle winds. Voles, ground squirrels, and birds can also

Arctic gentian in Rocky Mountain National Park.

disperse the seeds. Given the vagaries of weather in the alpine zone, many plant seeds can remain dormant and wait out unsuitably cool summers until a good year occurs.

Plants can also forgo sexual reproduction and multiply vegetatively. Some species spread by rhizomes, or via stolons, which are aboveground branches that grow new plantlets at their tips—like strawberries. Species with rhizomes make up a greater proportion of alpine plant communities where the growing season is shorter. Some species can develop roots where trailing branches touch the ground. These strategies have the downside of creating genetically identical individuals, which may limit the plant's ability to adapt to changing conditions. Consequently, most species hedge their bets and employ both sexual and asexual reproduction.

Whatever their survival strategies, plants of the alpine tundra have to go through their summer growth and reproduction quickly. Freezing temperatures that limit photosynthesis and windblown snow crystals that blast along like bits of sandpaper return quickly to the tundra, so summer is a frenzied time for plants.

Make Hay while the Sun Shines

Animals of the alpine tundra must be just as hardy and well-adapted as the alpine plants in order to survive the cold and the dryness. Very few species can manage this. Among invertebrates, there are the bees and flies crucial for plant pollination. Ants, beetles, grasshoppers, spiders, and soil mites are also present. Butterflies come winging in for pollination, but most species lay their eggs in the slightly gentler climates of lower elevations.

Like human tourists in the national park, birds are mostly summer visitors from the lowlands. Horned larks, rosy finches, and white-crowned sparrows nest and breed on the tundra. Ravens, hawks, prairie falcons, gray and Steller's jays, and Clark's nutcrackers come up to the tundra for shorter periods. Only the white-tailed ptarmigan can live on the tundra year-round, although during the winter, they prefer clumps of willows at or below tree line.

Ptarmigans are beautiful and beautifully adapted. Like many animals that need to withstand extreme cold, they are fat. The layer of fat, along with thick feathers, helps the bird retain body heat, and a ptarmigan can remain still for long periods of time, further retaining heat.

As with plants, summer is the time of activity and growth for ptarmigan. The birds breed during the short flowering season, and chicks are born precocial, ready to move about on their own and feed themselves. Chicks have

A ptarmigan hen along the Flattop Mountain Trail in Rocky Mountain National Park.

but a short time to grow before winter arrives. Consequently, they must learn to eat the high-protein plants that best support quick growth. The hens call their chicks to these foods by vocalizing and "tidbitting"—pecking at plants in a manner that stimulates the chicks to join their mother in eating the preferred plants.

Ptarmigan eat a wide variety of plants. The green leaves and flowers of mountain dryad, cinquefoil, buttercup, sedges, clovers, and snowball saxifrage are all palatable, but dwarf alpine willow and alpine bistort make up more than half of a ptarmigan's diet. As autumn begins, the birds still eat willow leaves, buds, and twigs most often, along with bistort, clovers, and mouse-ear before these plants are buried beneath the snow. Willow buds and twigs are the most abundant winter food, although mountain dryad and alder are also eaten.

During the long, cold winter nights, ptarmigan burrow into the snow to conserve body warmth and reduce exposure to the wind. Snow is so important to them that the birds choose sites with deep snow, into which they burrow as little as three inches or as deep as eleven inches. Paradoxically, warm winters are a problem. Warm winters mean shallower snow, but temperatures remain cold enough to limit ptarmigan survival. Relatively warm, clear, and calm weather during winter inhibits foraging by the birds, which in turn re-

A ptarmigan chick in August along the Flattop Mountain Trail, Rocky Mountain National Park.

duces their fat levels at the start of the breeding season in a manner that limits breeding success.

Ptarmigan populations are naturally widely dispersed among patches of suitable habitat. As tree line rises and tundra disappears under warming climate, ptarmigan habitat will become further fragmented, reducing gene exchange among dispersed groups and the ability of individual birds to move to more suitable habitat during periods of stress. As air temperatures have increased during recent decades, numbers of ptarmigan have declined to the point that scientific models suggest that the species could go extinct in Rocky Mountain National Park as climate continues to warm.

Mammals in the alpine zone also include visitors that mostly live at lower elevations, such as elk, bighorn sheep, coyotes, mule deer, bobcats, and mountain lions. Small mammals that manage to survive in the tundra year-round include pocket gophers, bushy-tailed woodrats, chipmunks, deer mice, golden-mantled ground squirrels, voles, and the photogenic pikas and marmots.

Pikas and marmots have different survival strategies. Pikas, which are members of the rabbit family, have hot little bodies. A high basal metabolic rate and good insulation keeps a pika's average body temperature at 104°F. This is a couple of degrees warmer than the body temperatures of the neighbors—ground squirrels, least chipmunks, and marmots. A hot body with good insulation helps pikas survive winters when the only food is whatever the animal has stored. Pikas remain active throughout the year and spend much of the summer harvesting plants that they dry in miniature hay piles and then stash in burrows. The pikas are thrifty: they can also eat their own fecal pellets to obtain the maximum food value from the plants that they have partly digested and excreted. Pikas obtain water from plants and conserve this water by excreting almost crystalline uric acids.

A hot body is not so helpful in a warming climate. Pikas have limited ability to survive overheating. When exposed to temperatures above 82°F for two hours, a pika dies. This is not usually a great threat for an animal that lives on alpine talus slopes. If daytime summer temperatures go too high, pikas can spend less time out in the sun or move a little more slowly. Careful human observations indicate that pikas commonly bustle about the surface for less than three minutes during hot weather before retreating to the coolness of a burrow in the rocks. The animal does have to make hay while the sun shines, however, and winter survival depends on having laid in sufficient supplies of food.

A pika collecting vegetation for its hay pile, Rocky Mountain National Park.

Several national parks in the western United States are participating in a project known as Pikas in Peril. Pikas are disappearing; population density is declining, the area occupied by the species is retreating to higher elevations, and, in some parks in the Great Basin, pikas have gone extinct. Warmer air temperatures are an obvious cause, but the details matter. Two of the most important details seem to be dryness and soil ice, as revealed by pika poop. Heat causes physiological stress in pikas, and that stress results in the formation of chemicals that can be measured in pika feces. Subsurface ice functions like a little air conditioner, keeping near-surface temperatures cooler in summer. Pikas living in sites with subsurface ice have lower concentrations of the telltale chemicals in their feces, evidence that they are less stressed by heat.

Dryness affects pika survival indirectly by controlling the amount and quality of food present. Pikas do best at sites with abundant and diverse flowering plants, rather than grasses. When summer gets hot and pikas must limit their foraging time to keep from overheating, they select the most nutritious and moist plants available. Microclimates become important. Well-shaded sites and east-facing slopes retain moisture better than exposed sites and west-facing slopes.

As heat and dryness increase, fewer animals will survive, and small populations will become more geographically and genetically isolated. Juvenile

Pika at the entrance to its burrow, Rocky Mountain National Park.

pikas disperse in late summer to early fall, and the little animals cannot travel long distances to find suitable new habitat.

During the last ice age, pikas were broadly distributed at lower elevations in the American west. As climate has warmed during the past ten thousand years, once-widespread populations have shrunk back to little pika islands in the tundra of the highest peaks. Now ecologists fear those islands may disappear in the rising tide of warmth. Thus far, pika strategies of selectively gathering moist, nutritious vegetation and limiting the length of their activities during hot summer days have prevented their extinction in Rocky Mountain National Park, but climate models predict that the animals are likely to disappear in the park. We will all be poorer for an alpine zone bereft of ptarmigan and pikas.

Marmots have a different strategy than hot-bodied pikas for surviving the rigors of the tundra. Marmots can hibernate for up to eight months, so they need to put on enough fat during the summer to fuel their slow winter metabolism. Young marmots have a faster metabolism than older animals, so the young ones may spend the winter sleeping with littermates in a bundle of marmot warmth.

The details of a burrow matter to a marmot. Each animal has a home bur-

Marmot among talus along the Flattop Mountain Trail, Rocky Mountain National Park.

row in which it normally spends the night and in which the young are raised. A home burrow always has at least three openings, and although a marmot may use another burrow as a temporary refuge when it cannot return to the home burrow, a marmot's behavior when another animal in its colony gives an alarm call indicates that there's no place like home. All the burrows in the colony, which consists of a male and his harem, are connected by trails that marmots have worn into the tundra. The prime locations are at the center of the colony and under a very large rock.

Back in the 1950s and early 1960s, zoologist Kenneth Armitage spent four summers scrutinizing the activity of a colony of marmots. His work is not as famous as the long-term studies of the charismatic apes (think Jane Goodall or Dian Fossey), but those careful observations revealed much about the life of a marmot outside its burrow.

A marmot's day has a regular routine that starts with emergence from the burrow when the first rays of sunlight reach the colony. Some of the marmots are early risers, others appear an hour later. On cloudy days, the marmots still emerge at about the same hour, perhaps prompted by internal needs: many marmots defecate immediately after they leave the burrow. That accomplished, a marmot gradually ramps up, with perhaps half an hour for

Portrait of a mature marmot, Rocky Mountain National Park.

grooming and sunning before the animal disperses to a feeding area and eats for a couple of hours. The rest of the day goes to digging, sunning, or resting in the burrow with occasional feeding forays. Young animals play together, and females play with their offspring. The whole colony takes a long siesta with minimal aboveground activity at midday, before coming out once more to feed until sunset. Activity in a marmot colony also generally drops when the air temperature rises above 68°F. All the marmots enter their burrows within thirty minutes after sunset, but each animal usually sits near the burrow entrance for a short time before entering.

A marmot typically lies down and crawls through vegetation while feeding, but marmot life is not all a bed of clover, and body language reveals much about the internal state of an animal. An adult going to the feeding grounds holds its tail up in a half-moon and waves it from side to side. When the alarm of a shrill whistle sounds, tails go down, and every marmot in the colony runs to the nearest burrow. A dominant animal holds the tail up and initiates grooming; a submissive animal holds the tail low, submits to grooming, or avoids dominant marmots. Two adults typically greet by sniffing each other's faces.

When overnight temperatures start to seriously drop, in late August or early September, the marmots enter hibernation. Pika survival during the winter depends on the hay that they have stockpiled. Marmot winter survival depends on the body fat they have accumulated. For all the small mammals of the tundra, the nutrition, moisture content, abundance, and distribution of the tundra plants are a matter of life and death.

Dust in the Wind

Their ability to photosynthesize can make plants seem self-sufficient, but of course they also require nutrients. Because dead plants fail to fully decompose in a cold climate, the nutrients stored in dead plant tissue remain there, rather than returning to the soil. Tundra plants have adapted to limited nutrients, but this is one of the limitations that slows their growth. Plants can compensate by retaining and recycling their own dead tissues and by intercepting windblown dust that contains vital nitrogen and phosphorus.

Just as windblown nitrogen is changing the plankton community of Loch Vale, so nitrogen raining down on the alpine tundra is changing the plant community. Botanist Bettie Willard began doing what she called "belly botany" on the tundra in Rocky Mountain National Park during the late 1950s. The name comes from the posture adopted by botanists engaged in surveying the tiny ground plants. Botanists have continued to survey plant communities since that time, and these surveys reveal the shift in community composition. Species of lichen that cannot withstand the newfound wealth of nitrogen are dying off. Grasses and herbaceous flowering plants such as clover, which can absorb the extra nitrogen, have increased in number. The changes are most pronounced where drifting snow that is rich in windblown dust provides a pulse of nitrogen when the snowdrifts melt.

The nitrogen-driven changes in plant communities mirror changes in soil microbes, and together these shifts increase the rates at which the element is cycled through the soil. The alpine soil is so thin, and the alpine plants have adapted to use so little nitrogen, that the alpine ecosystem has become saturated with this element during the past few decades. Now the tundra releases excess nitrogen from the soil to the air during the growing season and leaks it during snowmelt. Just as Loch Vale has a critical load, so the alpine tundra has a critical level of nitrogen inputs before the ecosystem starts to change. Estimated at about three pounds of nitrogen per acre per year for vegetation change and eight to thirteen pounds per acre per year for nitrate leaching from the tundra, these thresholds are lower than thresholds for most forest and shrublands ecosystems at lower elevations because the high-elevation ecosystem has so little capacity to absorb excess nitrogen.

Dust blowing into the alpine tundra brings other elements. Carbon also comes with the dust, but current levels of windblown carbon benefit the tundra. The soils of high-elevation talus slopes are among the most extreme environments for life in Rocky Mountain National Park. Even microbes have a hard time surviving in these soils. Few microbes mean few or no plants, which means coarse-textured soil with little organic matter, little ability to retain water, and few nutrients. Only about 10 percent of the windblown dust is carbon derived from dead plants in other regions, but this is enough to give the soil microbes and plants of the talus slopes a little boost. The dust also brings in clays and silts that increase the ability of the soil to retain water and further help the plants assimilate nutrients.

Across the tundra, atmospheric inputs of carbon in winter dust and dissolved carbon in summer rain are about one-third of the carbon made available by soil microbial communities. Tundra soil microbes are under snow for about nine months every year, limiting the time during which they can acquire carbon and nutrients. The microbes lose about as much carbon to the atmosphere as they create, which keeps tundra ecosystems carbon-limited and has kept scientists scratching their heads as to how these microbial communities can even obtain enough energy to sustain life. Part of the answer appears to be carbon carried in with dust. In dusty drylands, including the arid western United States, dust inputs subsidize the ecosystem by delivering carbon, nitrogen, phosphorus, iron, and calcium. Carbon, nitrogen, and phosphorus have in the past been in such limited supply that they limit plant growth and soil development. This limitation is now rapidly changing.

Glaciers and icefields have annual layers that record the inputs of windblown dust through time. These layers indicate that the Colorado Rocky Mountains have been getting about five to seven times more dust, mainly from the Colorado Plateau and the Great Basin (with a little bit of long-range transport from the deserts of Asia), since westward expansion of the United States in the mid-nineteenth century. Replacing native vegetation with croplands and grazing domesticated animals in the lowlands creates a signature of windblown dust in the highest of the highlands. The western United

Contrasting views of a snowfield just below the Continental Divide on the eastern side of the park, showing the effects of the 2006 dust deposition. (Views from Google Earth imagery).

States is not unique in this respect, however. Antarctic ice cores indicate a doubling of dust inputs from the nineteenth to the twentieth centuries as a result of global land use changes.

Dust makes snow melt. A dust covering absorbs sunlight and warmth to a much greater degree than do the reflective surfaces of snow and ice. Dust falling in midwinter can be buried by snow, but if the dust particles are large enough, the overlying snow melts and percolates below the dust. Dust falling in spring further darkens the snow surface and accelerates the warming and snowmelt. Dust in snow now shortens the duration of snow cover by several weeks in the enormous catchment of the Colorado River, creating a snowmelt runoff peak that is three weeks earlier and 5 percent smaller than historical averages at the long-term river-flow measurement site of Lee's Ferry, Arizona.

The Southern Rockies (northern New Mexico up to southern Wyoming), get more windblown dust than the Northern Rockies of Idaho and Montana because of the weather patterns that bring dust from deserts to the west and southwest into the Southern Rockies. These inputs appear to be accelerating. During the two decades between 1993 and 2014, winter and spring dust deposition increased by 81 percent in the Southern Rockies, making snowmelt earlier by seven to eighteen days. Some of this deposition includes black carbon, which results from incomplete combustion of fossil fuels in urban and industrial areas.

Individual storms now produce notable changes. Windstorms in Arizona, Utah, and western Colorado in mid-February 2006 created a dust cloud that blew across a wide swath from the San Juan Mountains in southern Colorado to the Medicine Bow Mountains of southern Wyoming. The storms deposited a layer of dust that remained visible for the rest of the winter. That layer was credited with both accelerating snowmelt, by eighteen to thirty-five days in different parts of the affected area that year, and increasing the occurrence of late-season avalanches in the Colorado Rockies.

A Conservative Ecosystem on the Edge

Except for the wind, the alpine tundra is a largely self-contained ecosystem. At lower elevations, streams receive nutrients from the adjacent forests and carry that material and energy downstream in a continuous, swirling exchange between terrestrial and aquatic organisms. Subalpine lakes receive nutrients gathered throughout their catchments and carried to the lake by snowmelt runoff. The alpine tundra has only the slow, slow weathering of rock and whatever the wind brings in. Consequently, the alpine plants have evolved to be conservative. They hold nutrients tightly by retaining their own dead tissues and trapping windblown dust. They store energy underground each summer to be ready for rapid growth as soon as the temperature rises slightly above freezing during the next growing season. They shelter their sensitive tissues, hugging the ground, clustering in cushions and mats, covering their surfaces in hairs, and keeping much of their mass underground. Their flowers follow the sun each day, little solar collectors tracking the source of warmth across the sky.

Streams are downstream flow with recycling, a prodigal system dependent on continual supply from the adjacent forests. The tundra is tight coils of recycling in place, a conservative system that until recently has released very few sediments, water droplets, or nutrients to adjacent lowlands. As excess nitrates deposited by the wind build up in the tundra, however, the composition of plant communities is changing, and the tundra is starting to leak nitrogen downslope to the subalpine and montane communities of the national park.

The tundra is also an ecosystem of contrasts. On clear days, views are limited only by the power of human eyesight, yet the old-growth ecosystem that rises only a few inches above the ground is as rich in details as a tropical rain forest. The terrain is relatively flat yet stands thousands of feet above sea level. The rock is strong, breaking down much more slowly than in adjacent lowlands, yet the barely visible wedges of delicate ice crystals slowly shatter the hard granite. The soil is fragile, so thin and vulnerable to wind erosion that the traffic of human feet can cause it to blow away. The plants are incredibly tough, living fast in summer and enduring the long winters, yet many of them cannot adapt to the recent bonanza of nitrogen. Small mammals like pikas and marmots adapt to subzero winter temperatures yet cannot withstand what a human would likely consider a pleasantly warm summer day. As air temperatures warm and the winds bring in more and more nitrogen, the increasing volume and speed of the hidden flows of the tundra leave the ecosystem balanced precariously on the edge of change.

Ecosystem Vital Signs

I have known my friend Sara for more than thirty years. I can recognize her at a distance by her walk and the general shape of her body. I can predict what she will find humorous or intriguing, what will make her angry or give her great pleasure. I understand something of the essence of Sara as an individual, yet I have little understanding of how she, as an organism, functions on a daily basis. I do not know the details of her genetic inheritance, let alone the molecular and cellular biology and intricate biochemical reactions that keep her alive and acting as a distinctive individual. I know some of her personal history, but I cannot pretend to understand her entire lifetime of experiences and how she perceives them. In other words, at some level Sara is a mystery to me, as I am to her, and to myself. Obviously, we can each get along reasonably well understanding what are in a sense the superficial details of ourselves and the individuals around us. We can see when something is obviously wrong with a friend—she is limping, looks tired, or is gaining or losing too much weight—but many times we cannot detect the subtle inner ills that eventually escalate into obvious problems. When problems do manifest, we need to consult an expert in some aspect of medical or psychological science.

Similarly, I can see the outer details of a subalpine forest, a mountain stream or lake, or a beaver meadow. I can watch the environment through time, observing seasonal changes and, over years, perhaps noting progressive changes such as beetle-caused death of trees or changes in the clarity of the lake water. I am likely to notice more details the longer and more carefully I look, but there will always be fundamental details that I do not see. I cannot see the chemicals passing between organisms in the rhizosphere, the invisible rain of nitrates, or how the epiphytes in the forest canopy absorb nutrients from the passing air. At most, I can see the effects of these processes once they change enough to cause more obvious responses in the environment, such as gradual takeover of the alpine tundra by plants that can thrive with extra nitrogen.

Like medical doctors, environmental scientists must measure invisible flows and observe the ecosystem changes caused by variations in these flows in order to understand how the ecosystem functions. All natural systems vary through time. A blowdown kills the trees in a portion of forest, and different species germinate in the newly created opening over the decade that follows. A year with a large snowmelt flood removes some of the logjams that have accumulated along a subalpine stream during preceding years; several years pass before newly formed logjams replace them.

These fluctuations through time define what ecologists call the natural range of variability. This range is the upper and lower limits of some measurable aspect of an ecosystem—the pH of lake water, the abundance of phytoplankton, the number of species of songbirds—in the absence of human manipulation. In an era of human-induced warming climate, the tricky part is finding an ecosystem that is not strongly influenced by deliberate or inadvertent human manipulation. Natural records such as lake sediments help define the range of variability present prior to intensive human alteration, as does careful study of the least-altered ecosystems. Individual examples of an ecosystem such as a subalpine lake differ from one another in sometimes subtle ways. A lake formed on granite bedrock might have different chemistry than a lake formed on the type of rock known as schist. Only through painstaking observations and measurements can scientists understand the natural range of variability for a particular ecosystem. And only through such measurements can they hope to document or predict when human-caused changes in an ecosystem have gone beyond the natural range and pushed the ecosystem toward the tipping point of an irreversible change.

Scientists continually measure the vital signs of an ecosystem. Just as a medical doctor routinely measures the vital signs of body temperature, pulse rate, respiration rate, and blood pressure to detect and monitor medical problems in an individual, so scientists try to measure characteristics that can reveal the health of an ecosystem. These measurements are a community effort, with each scientist contributing a particular expertise and insights. They are also very much the effect of passionate engagement with what might seem to other people to be trivial minutiae. Yet it is largely because a core group of scientists has measured water chemistry every week for thirty years at Loch Vale that we know how and why subalpine lakes in the Colorado Front Range are changing.

The something hidden within the ranges is there for each of us to find. From the intricate markings of the interior of an arctic gentian flower to the fingernail-sized patch of lichen high up the trunk of a ponderosa pine, the landscape of Rocky Mountain National Park rewards those who take the time to look carefully and closely. Returning to the same portions of the landscape through the seasons and the years reveals ongoing changes. A tree fallen into a stream collects smaller pieces and forms a logjam spanning the channel and backing up a deep pool rich in fallen leaves. A rockfall strips a patch of cliff to pale new rock that gradually acquires lichens and the stain of chemical reactions from snowmelt and rainwater flowing across the rock. These visible changes, like the chemical tracers in water, the remains of long-dead phytoplankton in lake sediments, or the historical records of plant abundance on the tundra, reveal that the landscape of the national park is continually altering. The downstream flow of water makes it seem intuitive that you can never step in the same river twice, yet the largely invisible flows that underpin every inch of the park suggest that you can also never step twice in the same forest, or meadow, or lake. Just as a person—or any living organism—changes throughout her lifespan, so does a landscape.

If there is any single insight to be drawn from the knowledge I have summarized in this book, it is that ecosystems are like icebergs: what we directly perceive is only a small portion of the total flows of energy and materials that sustain ecosystems. Fundamentally, we cannot afford to ignore what we cannot directly perceive. There is a great deal hidden in the ranges.

Many writers have quoted Charles Darwin's famous lines, "There is grandeur in this view of life." I join them. There is grandeur in this view of life, in which the invisible underpinnings of existence are revealed to be far more intricate, interdependent, and independent of us and our gods than we might have imagined. We will never find all that is hidden in the ranges. That remains our challenge and our salvation.

BIBLIOGRAPHY

General

Armstrong, D. M., J. P. Fitzgerald, and C. A. Meaney. 2011. *Mammals of Colorado.* 2nd ed. Boulder: University Press of Colorado.

Culver, D. R., and J. M. Lemly. 2013. *Field Guide to Colorado's Wetland Plants: Identification, Ecology and Conservation.* Fort Collins: Colorado Natural Heritage Program.

Duft, J. F., and R. K. Moseley. 1989. *Alpine Wildflowers of the Rocky Mountains.* Missoula, MT: Mountain Press.

Shaw, R. B. 2008. *Grasses of Colorado.* Boulder: University Press of Colorado.

Spellenberg, R., C. J. Earle, and G. Nelson. 2014. *Trees of Western North America.* Princeton, NJ: Princeton University Press.

Strickler, D. 1988. *Forest Wildflowers.* Columbia Falls, MT: The Flower Press.

Strickler, D. 1990. *Alpine Wildflowers.* Columbia Falls, MT: The Flower Press.

Ward, J. V., B. C. Kondratieff, and R. E. Zuellig. 2002. *An Illustrated Guide to the Mountain Stream Insects of Colorado.* 2nd ed. Boulder: University Press of Colorado.

Weber, W. A. 1976. *Rocky Mountain Flora.* Niwot: University Press of Colorado.

Willard, B. E., M. T. Smithson. nd. *Alpine Wildflowers of the Rocky Mountains.* Estes Park, CO: Rocky Mountain Nature Association.

Chapter 1: Montane Forest Hillslope

Baldocchi, D. 2008. "Breathing" of the Terrestrial Biosphere: Lessons Learned from a Global Network of Carbon Dioxide Flux Measurement Systems. *Australian Journal of Botany* 56:1–26.

Beiler, K. J., D. M. Durall, S. W. Simard, S. A. Maxwell, and A. M. Kretzer. 2010. Architecture of the Wood-Wide Web: *Rhizopogon* spp. Genets Link Multiple Douglas-Fir Cohorts. *New Phytologist* 185:543–553.

Datta, R., and S. Paul. 2016. Wood Wide Web. *Science Reporter* (April): 42–43. file:///C:/Users/User/Desktop/Datta%20&%20Paul.pdf.

Edburg, S. L., J. A. Hicke, P. D. Brooks, E. G. Pendall, B. E. Ewers, U. Norton, D. Gochis, E. D. Gutmann, and A. J. H. Meddens. 2012. Cascading Impacts of Bark Beetle-Caused Tree Mortality on Coupled Biogeophysical and Biogeochemical Processes. *Frontiers in Ecology and Environment* 10:416–424.

Franklin, J. F., H. H. Shugart, and M. E. Harmon. 1987. Tree Death as an Ecological Process. *BioScience* 37:550–556.

Hicke, J. A., M. C. Johnson, J. L. Hayes, and H. K. Preisler. 2012. Effects of Bark Beetle-Caused Tree Mortality on Wildfire. *Forest Ecology and Management* 271:81–90.

LaMalfa, E. M. 2007. Comparison of Water Dynamics in Aspen and Conifer: Implications for Ecology Water Yield Augmentation. MS thesis, Utah State University. All Graduate Theses and Dissertations. 6603. https://digitalcommons.usu.edu/etd/6603.

Molina, R., and M. Amaranthus. Rhizosphere Biology: Ecological Linkages between Soil Processes, Plant Growth, and Community Dynamics. In *Proceedings: Management and Productivity of Western-Montane Forest Soils*, edited by A. E. Harvey and L. F. Neuenschwander, 51–58. General Technical Report INT 280. Ogden, UT: USDA Forest Service.

Chapter 2: Stream Swirls

Brown, R. S., and W. C. Mackay. 1995. Fall and Winter Movements of and Habitat Use by Cutthroat Trout in the Ram River, Alberta. *Transactions of the American Fisheries Society* 124:873–885.

Coleman, M. A., and M. D. Fausch. 2007. Cold Summer Temperature Limits Recruitment of Age-0 Cutthroat Trout in High-Elevation Colorado Streams. *Transactions of the American Fisheries Society* 136:1231–1244.

Cunjak, R. A., and G. Power. 1986. Winter Habitat Utilization by Stream Resident Brook Trout (*Salvelinus fontinalis*) and Brown Trout (*Salmo trutta*). *Canadian Journal of Fisheries and Aquatic Sciences* 113:1970–1981.

Flemming, H.-C., and J. Wingender. 2010. The Biofilm Matrix. *Nature Reviews Microbiology* 8:623–633.

Hickman, T., and R. F. Raleigh. 1982. *Habitat Suitability Index Models: Cutthroat Trout*. Fort Collins, CO: US Fish and Wildlife Service.

Jakober, M. J., T. E. McMahon, R. F. Thurow, and C. G. Clancy. 1998. Role of Stream Ice on Fall and Winter Movements and Habitat Use by Bull Trout and

Cutthroat Trout in Montana Headwater Streams. *Transactions of the American Fisheries Society* 127:223–235.

Jakober, M. J., T. E. McMahon, and R. F. Thurow. 2000. Diel Habitat Partitioning by Bull Charr and Cutthroat Trout during Fall and Winter in Rocky Mountain Streams. *Environmental Biology of Fishes* 59:79–89.

Kempema, E. W., and R. Ettema. 2011. Anchor Ice Rafting: Observations from the Laramie River. *River Research and Applications* 27:1126–1135.

Kennedy, B. M., D. P. Peterson, and K. D. Fausch. 2003. Different Life Histories of Brook Trout Populations Invading Mid-Elevation and High-Elevation Cutthroat Trout Streams in Colorado. *Western North American Naturalist* 63:215–223.

Martin, S. 1981. Frazil Ice in Rivers and Oceans. *Annual Review of Fluid Mechanics* 13:379–397.

Pottinger, T. G., M. Rand-Weaver, and J. P. Sumpter. 2003. Overwinter Fasting and Re-feeding in Rainbow Trout: Plasma Growth Hormone and Cortisol Levels in Relation to Energy Mobilization. *Comparative Biology and Physiology* Part B 136:403–417.

Prowse, T. D. 2001. River-Ice Ecology. II: Biological Aspects. *Journal of Cold Regions Engineering* 15:17–33.

Ward, J. V. 1986. Altitudinal Zonation in a Rocky Mountain Stream. *Archives for Hydrobiology* (Supplement 74) 2:133–199.

Ward, J. V. 1992. A Mountain River. In *The Rivers Handbook: Hydrological and Ecological Principles*, vol. 1, edited by P. Calow and G. E. Petts, 493–510. Oxford: Blackwell Scientific.

Ward, J. V. 1994. Ecology of Alpine Streams. *Freshwater Biology* 32:277–294.

Ward, J. V., B. C. Kondratieff, and R. E. Zuellig. 2002. *An Illustrated Guide to Mountain Stream Insects of Colorado*. 2nd ed. Boulder: University Press of Colorado.

Chapter 3: Beaver Meadow

Costanzo, J. P., M. C. F. do Amaral, A. J. Rosendale, and R. E. Lee. 2013. Hibernation Physiology, Freezing Adaptation and Extreme Freeze Tolerance in a Northern Population of the Wood Frog. *Journal of Experimental Biology* 216:3461–3473.

Dinsmore, S. C., and D. L. Swanson. 2008. Temporal Patterns of Tissue Glycogen, Glucose, and Glycogen Phosphorylase Activity prior to Hibernation in Freeze-Tolerant Chorus Frogs, *Pseudacris triseriata*. *Canadian Journal of Zoology* 86:1095–1100.

Havens, K. J. 1996. *Plant Adaptations to Saturated Soils and the Formation of Hypertrophied Lenticels and Adventitious Roots in Woody Species.* Wetlands Program Technical Report no. 96-2. Williamsburg: Virginia Institute of Marine Science, College of William and Mary. http://publish.wm.edu/reports/640.

Higgins, S. A., and D. L. Swanson. 2013. Urea Is Not a Universal Cryoprotectant among Hibernating Anurans: Evidence from the Freeze-Tolerant Boreal Chorus Frog (*Pseudacris maculata*). *Comparative Biochemistry and Physiology, Part A* 164:344–350.

Hoag, J. C., N. Melvin, and D. Tilley. 2007. *Wetland Plants: Their Function, Adaptation and Relationship to Water Levels.* Riparian/Wetland Project Information Series No. 21. Washington, DC: USDA.

Hood, G. A., and S. E. Bayley. 2008. Beaver (*Castor canadensis*) Mitigate the Effects of Climate on the Area of Open Water in Boreal Wetlands in Western Canada. *Biological Conservation* 141:556–567.

Johnston, C. A. 2014. Beaver Pond Effects on Carbon Storage in Soils. *Geoderma* 213:371–378.

Kozlowski, T. T. 2002. Physiological-Ecological Impacts of Flooding on Riparian Forest Ecosystems. *Wetlands* 22:550–561.

Laurel, D., and E. Wohl. 2019. The Persistence of Beaver-Induced Geomorphic Heterogeneity and Organic Carbon Stock in River Corridors. *Earth Surface Processes and Landforms* 44:342–353.

Layne, J. R., and R. E. Lee. 1995. Adaptations of Frogs to Survive Freezing. *Climate Research* 5:53–59.

Merritt, D. M., and E. E. Wohl. 2002. Processes Governing Hydrochory along Rivers: Hydraulics, Hydrology, and Dispersal Phenology. *Ecological Applications* 12:1071–1087.

Merritt, D. M., and E. E. Wohl. 2006. Plant Dispersal along Rivers Fragmented by Dams. *River Research and Applications* 22:1–26.

Muths, E., R. D. Scherer, S. M. Amburgey, T. Matthews, A. W. Spencer, and P. S. Corn. 2016.First Estimates of the Probability of Survival in a Small-Bodied, High-Elevation Frog (Boreal Chorus Frog, *Pseudacris maculata*), or How Historical Data Can Be Useful. *Canadian Journal of Zoology* 94:599–606.

Naiman, R. J., C. A. Johnston, and J. C. Kelley. 1988. Alteration of North American Streams by Beaver. *BioScience* 38:753–762.

Polvi, L. E., and E. Wohl. 2012. The Beaver Meadow Complex Revisited: The Role of Beavers in Post-Glacial Floodplain Development. *Earth Surface Processes and Landforms* 37:332–346.

Ryden, H. 1989. *Lily Pond: Four Years with a Family of Beavers*. New York: Lyons and Burford.

Storey, K. B., and J. M. Storey. 1987. Persistence of Freeze Tolerance in Terrestrially Hibernating Frogs after Spring Emergence. *Copeia* 1987:720–726.

Swinnen, K. R. R., N. K. Hughes, and H. Leirs. 2015. Beaver (*Castor fiber*) Activity Patterns in a Predator-Free Landscape: What Is Keeping Them in the Dark? *Mammalian Biology* 80:477–483.

Wegener, P., T. Covino, and E. Wohl. 2017. Beaver-Mediated Lateral Hydrologic Connectivity, Fluvial Carbon and Nutrient Flux, and Aquatic Ecosystem Metabolism. *Water Resources Research* 53:4606–4623.

Chapter 4: Old-Growth Subalpine Forest

Anonymous. 2014. *Quick Guide Series: Spruce Beetle*. Colorado State Forest Service, FM 2014-1.

Arthur, M. A., and T. J. Fahey. 1990. Mass and Nutrient Content of Decaying Boles in an Engelmann Spruce–Subalpine Fir Forest, Rocky Mountain National Park, Colorado. *Canadian Journal of Forest Research* 20:730–737.

Brown, J. K., E. D. Reinhardt, and K. A. Kramer. 2003. *Coarse Woody Debris: Managing Benefits and Fire Hazard in the Recovering Forest*. General Technical Report RMRS-GTR-105. Ogden, UT: USDA Forest Service Rocky Mountain Research Station.

Brown, P. M., W. D. Shepperd, S. A. Mata, and D. L. McClain. 1998. Longevity of Windthrown Logs in a Subalpine Forest of Central Colorado. *Canadian Journal of Forest Research* 28:932–936.

Buskirk, S. W., S. C. Forrest, M. G. Raphael, and H. J. Harlow. 1989. Winter Resting Site Ecology of Marten in the Central Rocky Mountains. *Journal of Wildlife Management* 53:191–196.

Carrell, A. A., D. L. Carper, and A. C. Frank. 2016. Subalpine Conifers in Different Geographical Locations Host Highly Similar Foliar Bacterial Endophyte Communities. *FEMS Microbiology Ecology* 92:fiw124. doi:10.1093/femsec/fiw124.

Dell, I. H., and T. S. Davis. 2019. Effects of Site Thermal Variation and Physiography on Flight Synchrony and Phenology of the North American Spruce Beetle (Coleoptera: Curculionidae, Scolytinae) and Associated Species in Colorado. *Environmental Entomology* 48:998–1011.

Ellis, C. J. 2012. Lichen Epiphyte Diversity: A Species, Community and Trait-Based Review. *Perspectives in Plant Ecology, Evolution and Systematics* 14:131–152.

Gauslaa, Y. 2014. Rain, Dew, and Humid Air as Drivers of Morphology, Function and Spatial Distribution in Epiphytic Lichens. *The Lichenologist* 46:1–16.

Gough, L. P. 1975. Cryptogam Distributions on *Pseudotsuga menziesii* and *Abies lasiocarpa* in the Front Range, Boulder County, Colorado. *The Bryologist* 78:124–145.

Harmon, M. E., J. F. Franklin, F. J. Swanson, P. Sollins, S. V. Gregory, J. D. Lattin, N. H. Anderson, S. P. Cline, N. G. Aumen, J. R. Sedell, G. W. Lienkaemper, K. Cromack, and K. W. Cummins. 1986. Ecology of Coarse Woody Debris in Temperate Ecosystems. *Advances in Ecological Research* 15:133–302.

Keinath, D. A., and G. D. Hayward. 2003. Red-Backed Vole (*Clethrionomys gapperi*) Response to Disturbance in Subalpine Forests: Use of Regenerating Patches. *Journal of Mammalogy* 84:956–966.

Kueppers, L. M., J. Southon, P. Baer, and J. Harte. 2004. Dead Wood Biomass and Turnover Time, Measured by Radiocarbon, along a Subalpine Elevation Gradient. *Oecologia* 141:641–651.

Luoma, J. R. 1999. *The Hidden Forest: The Biography of an Ecosystem.* New York: Henry Holt.

Maser, C., and J. M. Trappe. 1984. *The Seen and Unseen World of the Fallen Tree.* General Technical Report PNW-164. Portland, OR: USDA Forest Service.

Tomaszewski, T., R. L. Boyce, and H. Sievering. 2003. Canopy Uptake of Atmospheric Nitrogen and New Growth Nitrogen Requirement at a Colorado Subalpine Forest. *Canadian Journal of Forest Research* 33:2221–2227.

Van Stan, J. T., and T. G. Pypker. 2015. A Review and Evaluation of Forest Canopy Epiphyte Roles in the Partitioning and Chemical Alteration of Precipitation. *Science of the Total Environment* 536:813–824.

Chapter 5: Subalpine Lake

Baron, J., ed. 1992. *Biogeochemistry of a Subalpine Ecosystem: Loch Vale Watershed.* New York: Springer-Verlag.

Baron, J. S. 2006. Hindcasting Nitrogen Deposition to Determine an Ecological Critical Load. *Ecological Applications* 16:433–439.

Baron, J., and A. S. Denning. 1993. The Influence of Mountain Meteorology on Precipitation Chemistry at Low and High Elevations of the Colorado Front Range, USA. *Atmospheric Environment* 27A:2337–2349.

Baron, J. S., D. S. Ojima, E. A. Holland, and W. J. Parton. 1994. Analysis of Nitrogen Saturation Potential in Rocky Mountain Tundra and Forest: Implications for Aquatic Systems. *Biogeochemistry* 27:61–82.

Baron, J. S., H. M. Rueth, A. M. Wolfe, K. R. Nydick, E. J. Allstott, J. T. Minear, and B. Moraska. 2000. Ecosystem Responses to Nitrogen Deposition in the Colorado Front Range. *Ecosystems* 3:352–368.

Baron, J. S., T. M. Schmidt, and M. D. Harman. 2009. Climate-Induced Changes in High Elevation Stream Nitrate Dynamics. *Global Change Biology* 15:1777–1789.

Birks, S. J., and T. W. D. Edwards. 2009. Atmospheric Circulation Controls on Precipitation Isotope–Climate Relations in Western Canada. *Tellus B: Chemical and Physical Meteorology* 61:566–576.

Bowen, G. J., and J. Revenaugh. 2003. Interpolating the Isotopic Composition of Modern Meteoric Precipitation. *Water Resources Research* 39:1299. doi:10.1029/2003WR002086.

Campbell, D. H., D. W. Clow, G. P. Ingersoll, M. A. Mast, N. E. Spahr, and J. T Turk. 1995. Processes Controlling the Chemistry of Two Snowmelt-Dominated Streams in the Rocky Mountains. *Water Resources Research* 31:2811–2821.

Christensen, D. L., B. R. Herwig, D. E. Schindler, and S. R. Carpenter. 1996. Impacts of Lakeshore Residential Development on Coarse Woody Debris in North Temperate Lakes. *Ecological Applications* 6:1143–1149.

Greene, S., K. M. W. Anthony, D. Archer, A. Sepulveda-Jauregui, and K. Martinez-Cruz. 2014. Modeling the Impediment of Methane Ebullition Bubbles by Seasonal Lake Ice. *Biogeosciences* 11:6791–6811.

Guyette, R. P., and W. G. Cole. 1999. Age Characteristics of Coarse Woody Debris (*Pinus strobus*) in a Lake Littoral Zone. *Canadian Journal of Fisheries and Aquatic Sciences* 56:496–505.

Guyette, R. P., W. G. Cole, D. C. Dey, and R. M. Muzika. 2002. Perspectives on the Age and Distribution of Large Wood in Riparian Carbon Pools. *Canadian Journal of Fisheries and Aquatic Sciences* 59:578–585.

Helmus, M. R., and G. G. Sass. 2008. The Rapid Effects of a Whole-Lake Reduction of Coarse Woody Debris on Fish and Benthic Macroinvertebrates. *Freshwater Biology* 53:1423–1433.

Klaus, J., and J. J. McDonnell. 2013. Hydrograph Separation Using Stable Isotopes: Review and Evaluation. *Journal of Hydrology* 505:47–64.

Liu, Z., G. J. Bowen, and J. M. Welker. 2010. Atmospheric Circulation Is Reflected in Precipitation Isotope Gradients over the Conterminous United States. *Journal of Geophysical Research* 115:D22120. doi:10.1029/2010JD014175.

McKnight, D., M. Brenner, R. Smith, and J. Baron. 1986. *Seasonal Changes in Phytoplankton Populations and Related Chemical and Physical Characteristics in Lakes in Loch Vale, Rocky Mountain National Park, Colorado.* Water-Resources Investigations Report 86-4101. Denver, CO: US Geological Survey.

Schindler, D. W., and B. R. Parker. 2002. Biological Pollutants: Alien Fishes in Mountain Lakes. *Water, Air, and Soil Pollution* 2:379–397.

Sueker, J. K., J. N. Ryan, C. Kendall, and R. D. Jarrett. 2000. Determination of Hydrologic Pathways during Snowmelt for Alpine/Subalpine Basins, Rocky Mountain National Park, Colorado. *Water Resources Research* 36:63–75.

Vadeboncoeur, Y., and D. M. Lodge. 2000. Periphyton Production on Wood and Sediment: Substratum-Specific Response to Laboratory and Whole-Lake Nutrient Manipulations. *Journal of the North American Benthological Society* 19:68–81.

Chapter 6: Alpine Flats

Allen, T., and J. A. Clarke. 2005. Social Learning of Food Preferences by White-Tailed Ptarmigan Chicks. *Animal Behaviour* 70:305–310.

Anderson, R. S. 1998. Near-Surface Thermal Profiles in Alpine Bedrock: Implications for the Frost Weathering of Rock. *Arctic and Alpine Research* 30:362–372.

Anderson, R. S., S. P. Anderson, and G. E. Tucker. 2013. Rock Damage and Regolith Transport by Frost: An Example of Climate Modulation of the Geomorphology of the Critical Zone. *Earth Surface Processes and Landforms* 38:299–316.

Anderson, R. S., C. A. Riihimaki, E. B. Safran, and K. R. MacGregor. 2006. Facing Reality: Late Cenozoic Evolution of Smooth Peaks, Glacially Ornamented Valleys, and Deep River Gorges of Colorado's Front Range. In *Tectonics, Climate, and Landscape Evolution*, edited by S. D. Willett, N. Hovius, M. T. Brandom, and D. M. Fisher, 397–418. Special Paper 398. Boulder, CO: Geological Society of America.

Anderson, S. P. 2019. Breaking It Down: Mechanical Processes in the Weathering Engine. *Elements* 15:247–252.

Anderson, S. P., R. S. Anderson, E. S. Hinckley, P. Kelly, and A. Blum. 2011. Exploring Weathering and Regolith Transport Controls on Critical Zone Development with Models and Natural Experiments. *Applied Geochemistry* 26:53–55.

Armitage, K. B. 1962. Social Behavior of a Colony of the Yellow-Bellied Marmot (*Marmota flaviventris*). *Animal Behaviour* 10:319–331.

Bowman, W. D., J. Murgel, T. Blett, and E. Porter. 2012. Nitrogen Critical Loads for Alpine Vegetation and Soils in Rocky Mountain National Park. *Journal of Environmental Management* 103:165–171.

Brown, J. L., and L. L. Knowles. 2012. Spatially Explicit Models of Dynamic

Histories: Examination of the Genetic Consequences of Pleistocene Glaciation and Recent Climate Change on the American Pika. *Molecular Ecology* 21:3757–3775.

Castillo, J. A., C. W. Epps, A. R. Davis, and S. A. Cushman. 2014. Landscape Effects on Gene Flow for a Climate-Sensitive Montane Species, the American Pika. *Molecular Ecology* 23:843–856.

Clow, D. W., G. P. Ingersoll, M. A. Mast, J. T. Turk, and D. H. Campbell. 2002. Comparison of Snowpack and Winter Wet-Deposition Chemistry in the Rocky Mountains, USA: Implications for Winter Dry Deposition. *Atmospheric Environment* 36:2337–2348.

Clow, D. W., M. W. Williams, and P. F. Schuster. 2016. Increasing Aeolian Dust Deposition to Snowpacks in the Rocky Mountains Inferred from Snowpack, Wet Deposition, and Aerosol Chemistry. *Atmospheric Environment* 146:183–194.

Eppes, M. C., and R. Keanini. 2017. Mechanical Weathering and Rock Erosion by Climate-Dependent Subcritical Cracking. *Reviews of Geophysics* 55:470–508.

Erb, L. P., C. Ray, and R. Guralnick. 2011. On the Generality of a Climate-Mediated Shift in the Distribution of the American Pika (*Ochotona princeps*). *Ecology* 92:1730–1735.

Erb, L. P., C. Ray, and R. Guralnick. 2014. Determinants of Pika Population Density vs. Occupancy in the Southern Rocky Mountains. *Ecological Applications* 24:429–435.

Galbreath, K. E., D. J. Hafner, and K. R. Zamudio. 2009. When Cold Is Better: Climate-Driven Elevation Shifts Yield Complex Patterns of Diversification and Demography in an Alpine Specialist (American Pika, *Ochotona princeps*). *Evolution* 63:2848–2863.

Garber, J. 2013. Using in situ Cosmogenic Radionuclides to Constrain Millennial Scale Denudation Rates and Chemical Weathering Rates on the Colorado Front Range. Unpublished MS thesis, Colorado State University, Fort Collins.

Jeffress, M. R., T. J. Rodhouse, C. Ray, S. Wolff, and C. W. Epps. 2013. The Idiosyncrasies of Place: Geographic Variation in the Climate-Distribution Relationships of the American Pika. *Ecological Applications* 23:864–878.

Knight, J., and S. W. Grab. 2014. Lightning as a Geomorphic Agent on Mountain Summits: Evidence from Southern Africa. *Geomorphology* 204:61–70.

Lawrence, C. R., J. C. Neff, and G. L. Farmer. 2011. The Accretion of Aeolian Dust in Soils of the San Juan Mountains, Colorado, USA. *Journal of Geophysical Research* 116:F02013.

Lawrence, C. R., T. H. Painter, C. C. Landry, and J. C. Neff. 2010. Contemporary

Geochemical Composition and Flux of Aeolian Dust to the San Juan Mountains, Colorado, United States. *Journal of Geophysical Research* 115:G03007.

Ley, R. E., M. W. Williams, and S. K. Schmidt. 2004. Microbial Population Dynamics in an Extreme Environment: Controlling Factors in Talus Soils at 3750 m in the Colorado Rocky Mountains. *Biogeochemistry* 68:313–335.

MacArthur, R. A., and L. C. H. Wang. 1973. Physiology of Thermoregulation in the Pika, *Ochotona princeps. Canadian Journal of Zoology* 51:11–16.

MacArthur, R. A., and L. C. H. Wang. 1974. Behavioral Thermoregulation in the Pika *Ochotona princeps*: A Field Study Using Radiotelemetry. *Canadian Journal of Zoology* 52:353–358.

May, T. A., and C. E. Braun. 1972. Seasonal Foods of Adult White-Tailed Ptarmigan in Colorado. *Journal of Wildlife Management* 36:1180–1186.

Mladenov, N., M. W. Williams, S. K. Schmidt, and K. Cawley. 2012. Atmospheric Deposition as a Source of Carbon and Nutrients to an Alpine Catchment of the Colorado Rocky Mountains. *Biogeosciences* 9:3337–3355.

Painter, T. H., S. M. Skiles, J. S. Deems, A. C. Bryant, and C. C. Landry. 2012. Dust Radiative Forcing in Snow of the Upper Colorado River Basin: 1. A 6 Year Record of Energy Balance, Radiation, and Dust Concentrations. *Water Resources Research* 48:W07521.

Pielou, E. C. 1994. *A Naturalist's Guide to the Arctic*. Chicago: University of Chicago Press.

Rhoades, C., K. Elder, and E. Greene. 2010. The Influence of an Extensive Dust Event on Snow Chemistry in the Southern Rocky Mountains. *Arctic, Antarctic, and Alpine Research* 42:98–105.

Skiles, S. M., and T. Painter. 2017. Daily Evolution in Dust and Black Carbon Content, Snow Grain Size, and Snow Albedo during Snowmelt, Rocky Mountains, Colorado. *Journal of Glaciology* 63:118–132.

Walker, J. J., and N. R. Pace. 2007. Endolithic Microbial Ecosystems. *Annual Reviews of Microbiology* 61:331–347.

Wang, G., N. T. Hobbs, K. M. Giesen, H. Galbraith, D. S. Ojima, and C. E. Braun. 2002. Relationships between Climate and Population Dynamics of White-Tailed Ptarmigan *Lagopus leucurus* in Rocky Mountain National Park, Colorado, USA. *Climate Research* 23:81–87.

Wilkening, J. L., C. Ray, and J. Varner. 2015. Relating Sub-surface Ice Features to Physiological Stress in a Climate Sensitive Mammal, the American Pika (*Ochotona princeps*). *PLOS One* 10 (3): e0119327.

Willard, B. E., and J. W. Marr. 1970. Effects of Human Activities on Alpine Tundra Ecosystems in Rocky Mountain National Park, Colorado. *Biological Conservation* 2:257–265.

Willard, B. E., D. J. Cooper, and B. C. Forbes. 2007. Natural Regeneration of Alpine Tundra Vegetation after Human Trampling: A 42-Year Data Set from Rocky Mountain National Park, Colorado, USA. *Arctic, Antarctic, and Alpine Research* 39:177–183.

Wohl, E. 2016. *Rhythms of Change in Rocky Mountain National Park*. Lawrence: University Press of Kansas.

Yandow, L. H., A. D. Chalfoun, and D. F. Doak. 2015. Climate Tolerances and Habitat Requirements Jointly Shape the Elevational Distribution of the American Pika (*Ochotona princeps*), with Implications for Climate Change Effects. *PLOS One* 10:e0131082.

Zwinger, A. H., and B. E. Willard. 1996. *Land above the Trees: A Guide to American Alpine Tundra*. Boulder, CO: Johnson Books.

SCIENTIFIC NAMES

Birds
boreal owl (*Aegolius funereus*)
Wilson's warbler (*Cardellina pusilla*)
ouzel (*Cinclus mexicanus*)
olive-sided flycatcher (*Contopus cooperi*)
common raven (*Corvus corax*)
Steller's jay (*Cyanocitta stelleri*)
horned lark (*Eremophila alpestris*)
prairie falcon (*Falco mexicanus*)
MacGillivray's warbler (*Geothlypis tolmiei*)
white-tailed ptarmigan (*Lagopus leucurus*)
rosy finch (*Leucosticte atrata*)
Clark's nutcracker (*Nucifraga columbiana*)
gray jay (*Perisoreus canadensis*)
downy woodpecker (*Picoides pubescens*)
three-toed woodpecker (*Picoides tridactylus*)
hairy woodpecker (*Picoides villosus*)
mountain chickadee (*Poecile gambeli*)
broad-tailed hummingbird (*Selasphorus platycercus*)
rufous hummingbird (*Selasphorus rufus*)
yellow-rumped warbler (*Setophaga coronata*)
mountain bluebird (*Sialia currucoides*)
white-crowned sparrow (*Zonotrichia leucophrys*)

Fish
suckers (*Catostomus catostomus, C. commersonii*)

Colorado River cutthroat trout (*Oncorhynchus clarkii pleuriticus*)
greenback cutthroat trout (*Oncorhynchus clarkia stomias*)
rainbow trout (*Oncorhynchus mykiss*)
longnose dace (*Rhinichthys cataractae*)
brown trout (*Salmo trutta*)
brook trout (*Salvelinus fontinalis*)

Amphibians
tiger salamander (*Ambystoma tigrinum*)
boreal toad (*Bufo boreas*)
boreal chorus frog (*Pseudacris maculata*)
western chorus frog (*Pseudacris triseriata*)
northern leopard frog (*Rana pipiens*)
wood frog (*Rana sylvatica*)

Insects
mountain pine beetle (*Dendroctonus ponderosae*)
spruce beetle (*Dendroctonus rufipennis*)
mayflies (order Ephemeroptera)
St. Lawrence tiger moth (*Platarctia parthenos*)
stoneflies (order Plecoptera)
caddisflies (order Trichoptera)

Mammals
moose (*Alces alces*)
golden-mantled ground squirrel (*Callospermophilus lateralis*)
coyote (*Canis latrans*)
beaver (*Castor canadensis*)
elk (*Cervus elaphus*)
big brown bat (*Eptesicus fuscus*)
bobcat (*Felis rufus*)
silver-haired bat (*Lasionycteris noctivagans*)
hoary bat (*Lasiurus cinereus*)
snowshoe hare (*Lepus americanus*)
otter (*Lutra canadensis*)
Canada lynx (*Lynx canadensis*)
yellow-bellied marmot (*Marmota flaviventris*)
pine marten (*Martes americana*)

long-tailed vole (*Microtus longicaudus*)
montane vole (*Microtus montanus*)
ermine (*Mustela erminea*)
southern red-backed vole (*Myodes gapperi*)
little brown bat (*Myotis lucifugus*)
least chipmunk (*Neotamias minimus*)
bushy-tailed woodrat (*Neotoma cinerea*)
mink (*Neovison vison*)
pika (*Ochotona princeps*)
mule deer (*Odocoileus hemionus*)
bighorn sheep (*Ovis canadensis*)
deer mice (*Peromyscus maniculatus*)
mountain lion (*Puma concolor*)
Abert's squirrel (*Sciurus aberti*)
masked shrew (*Sorex cinereus*)
water shrew (*Sorex palustris*)
rabbit (*Sylvilagus floridanus*)
pine squirrel or chickaree (*Tamiasciurus hudsonicus*)
northern pocket gopher (*Thomomys talpoides*)
black bear (*Ursus americanus*)
red fox (*Vulpes vulpes*)

Herbaceous plants
baneberry (*Actaea rubra*)
angelica (*Angelica arguta*)
Colorado columbine (*Aquilegia coerulea*)
Rocky Mountain columbine (*Aquilegia saximontana*)
heartleaf arnica (*Arnica cordifolia*)
American bistort (*Bistorta bistortoides*)
alpine bistort (*Bistorta vivipara*)
mariposa lily (*Calochortus gunnisonii*)
marsh marigold (*Caltha palustris*)
sedges (*Carex* spp.)
paintbrush (*Castilleja rhexifolia* and *C. sulphurea* in the alpine)
mouse-ear (*Cerastium* spp.)
pipsissewa (*Chimaphila umbellata*)
queen's crown (*Clementsia rhodantha*)
spotted coralroot (*Corallorhiza maculata*)

shooting star (*Dodecatheon pulchellum*)
mountain dryad or mountain aven (*Dryas octopetala*)
arctic gentian (*Gentianodes algida*)
white bog orchid (*Habenaria dilatata*)
cow parsnip (*Heracleum lanatum*)
Rocky Mountain iris (*Iris missouriensis*)
common duckweed (*Lemna minor*)
lupine (*Lupinus* spp.)
Oregon grape (*Mahonia repens*)
elephanthead (*Pedicularis groenlandica*)
bistort (*Polygonum bistorta*)
cinquefoil (*Potentilla diversifolia*)
Parry's primrose (*Primula parryi*)
pinedrops (*Pterospora andromedea*)
buttercup (*Ranunculus* spp.)
king's crown (*Rhodiola integrifolia*)
wax currant (*Ribes cereum*)
wild rose (*Rosa woodsii*)
snowball saxifrage (*Saxifraga rhomboidea*)
twisted stalk (*Streptopus amplexifolius*)
clovers (*Trifolium* spp.)
blueberries (*Vaccinium* spp.)

Woody plants
subalpine fir (*Abies lasiocarpa*)
alder (*Alnus incana*)
cinquefoil (*Dasiphora fruticosa*)
common juniper (*Juniperus communis*)
Engelmann spruce (*Picea engelmannii*)
blue spruce (*Picea pungens*)
lodgepole pine (*Pinus contorta*)
limber pine (*Pinus flexilis*)
ponderosa pine (*Pinus ponderosae*)
aspen (*Populus tremuloides*)
Douglas-fir (*Pseudotsuga menziesii*)
antelope bitterbrush (*Purshia tridentata*)
dwarf alpine willow (*Salix reticulata*)
willows (*Salix* spp.)

Other

fly agaric (*Amanita muscaria*)
porcini (*Boletus edulis*)
blue stain fungus (*Grosmannia clavigera*)

INDEX